A LETTER FROM JACK DANGERMOND

Inside this book, you will find a small sampling of geographic information system (GIS) maps created by Esri® users and shared at our annual user conference in 2022.

"Mapping Common Ground" was the theme of this conference. These words were chosen carefully to describe the amazing efforts of GIS professionals in applying geographic information to help find solutions to problems affecting us all.

Our current lack of environmental understanding, together with our failure to collaborate, signals the need for broadly practicing the concepts of mapping common ground—to work on solutions that benefit all sectors.

Each of the maps included in this volume of *Esri Map Book* tells a story and, in many cases, illustrates how organizations have created a better understanding that helps them find creative solutions to complex problems.

These maps also clearly provide the evidence that GIS is expanding rapidly and delivering value to almost every discipline and type of organization. At the center of this expansion is the growing awareness of the power of maps and GIS for integrating and analyzing geographic data as a kind of enterprise platform. Implementations at the enterprise scale are impacting and transforming workflows and organizations everywhere. And they are bringing together solutions with a common purpose.

Part of the power of GIS involves the ability to easily integrate multiple types and layers of map information, including environmental factors, economic layers, and cultural and social information.

Beyond analytics and visualization, this platform is also providing a framework for people to design, collaborate, and find solutions—a foundation for positive action. We like to refer to this as "The Geographic Approach."

As more and more people use our software and The Geographic Approach becomes embedded in daily activities and business operations, I am hopeful we can continue to make progress in creating a smarter and more sustainable future.

Warm regards,

Jack Dangermond

CONTENTS

PETALUMA GAP VITICULTURE

HDR Engineering Inc.
Sacramento, California, USA
By Keir Keightley

CONTACT

Keir Keightley
keir.keightley@hdrinc.com

SOFTWARE

ArcGIS® Pro, ArcGIS Web AppBuilder

DATA SOURCES

Dept of Treasury/TTB, ArcGIS StreetMap™ Premium,
US Census Bureau, USGS

The Petaluma Gap in southern Sonoma County and
northern Marin County in California is known for a
consistent pattern of onshore winds. The topography and
proximity to the Pacific Ocean and San Francisco Bay make
it suitable for grape growing and wine making. The official
name of this area—the Petaluma Gap American Viticultural
Area (AVA)—was recently approved by the Federal Trade
and Tax Bureau for use by wine professionals in branding
their grapes and wines.

Courtesy of HDR Engineering Inc.

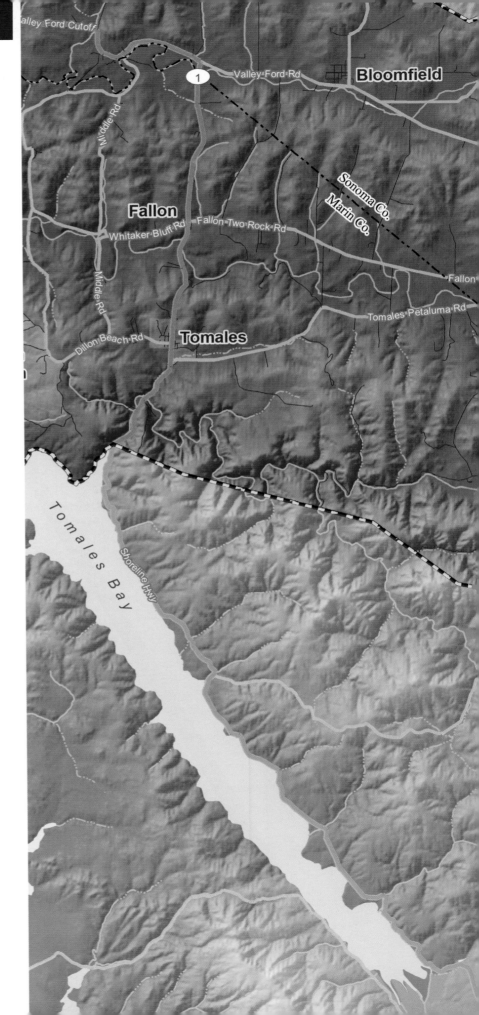

MANAGING AND SHARING GIS PROJECT DATA

Dawood Engineering
Pittsburgh,
Pennsylvania, USA
By Jodie Gosselin

When the Georgia-based energy services holding company Southern Company Gas (SCG) required a more efficient land management solution for its 100-mile Illinois pipeline project, the company turned to GIS. This project dashboard shows how a Microsoft Excel spreadsheet was geoenabled to manage the land acquisition process.

The GIS portal empowers the project team—executives, managers, land agents, and consultants—to access, integrate, analyze, and visualize easement acquisition information in real time and make informed business decisions.

This map provides a visual overview of proposed route details—such as survey permission, plat development, and acquisition. SCG's design engineering firm maintains the

core basemap layers whereas updated land acquisition data is pulled from the land portal. Land agents can filter and sort by project phase or landowner name. Project team members can view custom reports and download CSV files from this interface using query and reporting tools.

Courtesy of Dawood Engineering.

CONTACT
Jodie Gosselin
jodie.gosselin@dawood.net

SOFTWARE
ArcGIS Pro,
ArcGIS Online,
ArcGIS Dashboards,
ArcGIS StoryMaps℠,
ArcGIS Survey123

DATA SOURCES
Southern Company Gas,
Digital Map Products

E-COMMERCE CUSTOMER TRENDS

Esri Indonesia, Jakarta Selatan, Indonesia,
By Andika Hadi Hutama

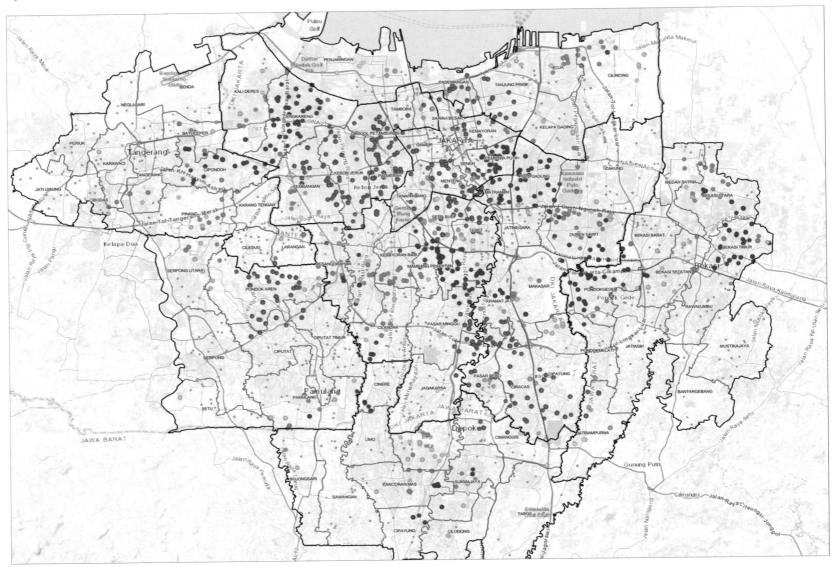

Courtesy of Rifqi Oktavianto.

As e-commerce business in Indonesia has grown, it has enabled the use of massive amounts of location-enabled purchasing data. Businesses can use location and temporal data to seek answers to questions such as: Where are our best customers? Where are the purchasing hot spots and cold spots? What are the chances of a random purchasing spike? Where are the emerging buying patterns occurring? At which locations and during what periods should we allocate more of our resources or products?

Approximately 30,000 data points of purchase in greater Jakarta were obtained and used to create this space-time cube. Using location intelligence, businesses can put descriptive and predictive analytics on a map and gain powerful insights.

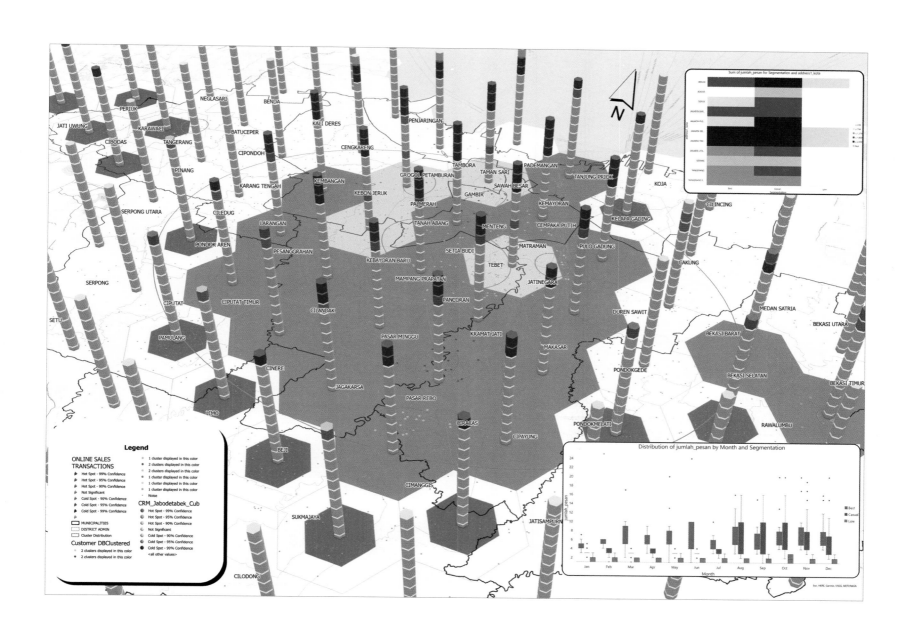

CONTACT
Andika Hadi Hutama
ahhutama@esriindonesia.co.id

SOFTWARE
ArcGIS Dashboards,
ArcGIS Pro,
ArcGIS StoryMaps,
ArcGIS Enterprise,
ArcGIS Online

DATA SOURCES
Esri Indonesia

COVID-19 HOUSEHOLD RELOCATIONS NEAR TORONTO

Environics Analytics Group Ltd.
Toronto, Canada
By Thomas Lillo, Michael McDuffee, Philip Tananka,
and Jesse Terhorst

CONTACT
Philip Tananka
philip.tananka@environicsanalytics.com

SOFTWARE
ArcGIS Pro, Altery, Snowflake

DATA SOURCES
WealthScapes, MobileScapes

This map shows Toronto, Ontario, residential relocation patterns in the year immediately after the COVID-19 pandemic, as well as changes to the cost of housing in those areas. Using privacy-compliant movement and financial data, this map explores the relationship between households leaving Toronto and the increases in housing costs across the province. Blue lines indicate the number of households that have moved, and red areas show the average cost of housing in the districts to which these households relocated.

Long-term changes in the common evening location of mobile devices were used to determine whether a device had moved out of the downtown core and into a surrounding municipality. This provided a map for investigating movement-related trends and informing business decisions related to real estate, logistics, and consumer behavior.

Courtesy of Environics Analytics Group Ltd.

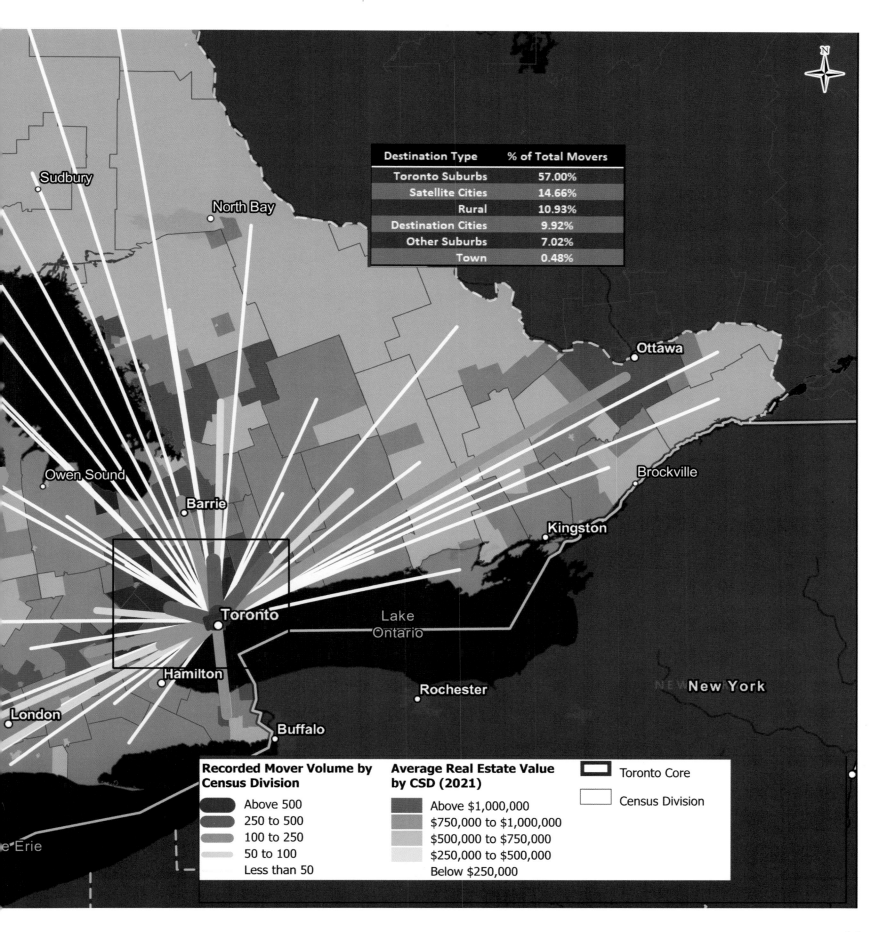

Destination Type	% of Total Movers
Toronto Suburbs	57.00%
Satellite Cities	14.66%
Rural	10.93%
Destination Cities	9.92%
Other Suburbs	7.02%
Town	0.48%

Recorded Mover Volume by Census Division
- Above 500
- 250 to 500
- 100 to 250
- 50 to 100
- Less than 50

Average Real Estate Value by CSD (2021)
- Above $1,000,000
- $750,000 to $1,000,000
- $500,000 to $750,000
- $250,000 to $500,000
- Below $250,000

- Toronto Core
- Census Division

CREDIT CARD USE PATTERNS

Retail Profit Management
San Diego, California, USA
By Steve Lackow

CONTACT
Steve Lackow
slackow@rpmconsulting.com

SOFTWARE
ArcMap, ArcGIS Business Analyst™

DATA SOURCES
RPM Consulting Inc.

Maxed Out identifies regions in San Diego, California, based on how close people are to reaching their credit card limit. Compared with the average nationwide line of credit use, a pattern emerges of high credit card use in the urban core and less reliance on credit in the surrounding areas. The Interstate 8 corridor is a major line of division: populations south of the interstate tend to be maxed out on their credit cards. Other areas of concern include El Cajon, Mission Beach, Ocean Beach, and the Golden Triangle (the largely residential area bordered by Interstate 5, Interstate 805, and Route 52).

Courtesy of Retail Profit Management.

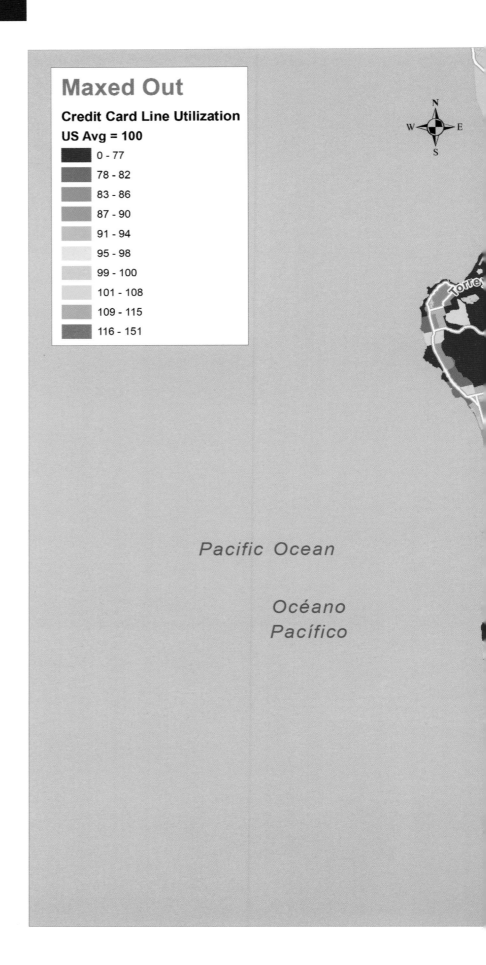

Maxed Out

Credit Card Line Utilization
US Avg = 100

- 0 - 77
- 78 - 82
- 83 - 86
- 87 - 90
- 91 - 94
- 95 - 98
- 99 - 100
- 101 - 108
- 109 - 115
- 116 - 151

Pacific Ocean

Océano Pacífico

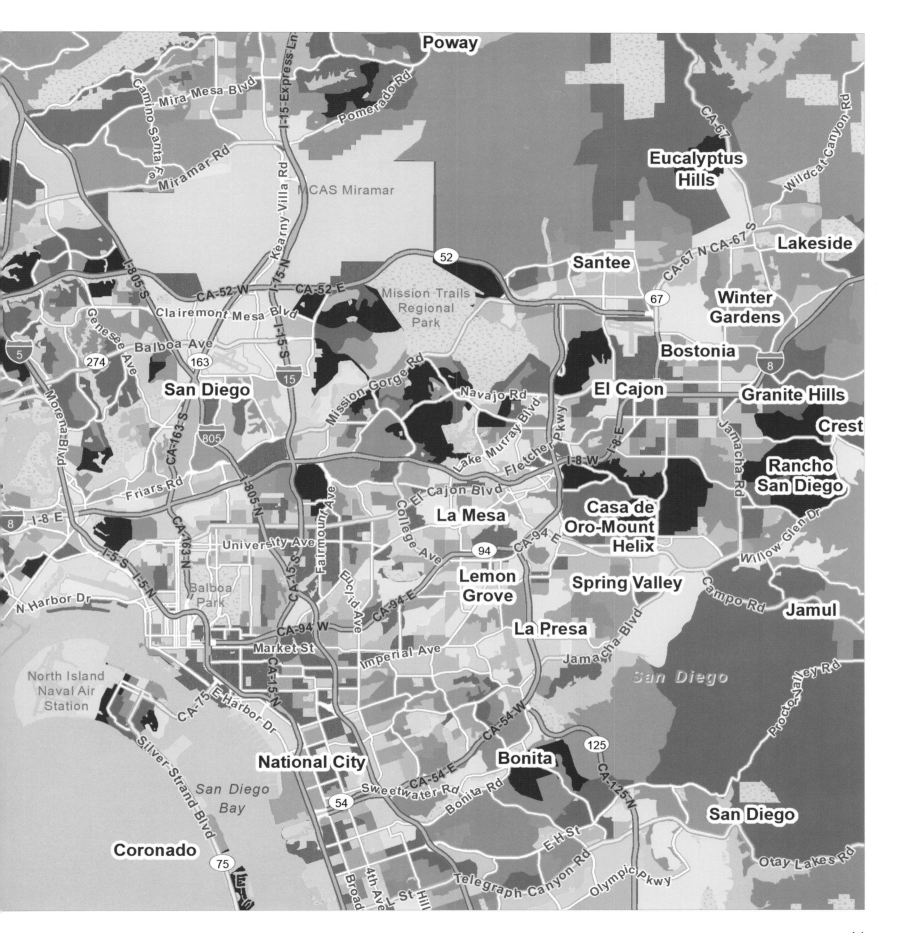

OLD CAIRO

Mai Ghaly Elgazzar
Dubai, United Arab Emirates
By Mai Ghaly Elgazzar

CONTACT
Mai Ghaly Elgazzar
mai.ghaly23@hotmail.com

SOFTWARE
ArcGIS Pro, ArcGIS Vector Tile Style Editor, Adobe PhotoShop

DATA SOURCES
ArcGIS Online, Google Maps

Egyptians call it Al-Qahirah ("The Victorious"), and some call it Qahirat el-Moez ("The City of a Thousand Minarets"). The original name of Cairo was given by Jawhar al-Siqilli, who began construction in 969 CE, designing it as the capital city of the Fatimid Caliphate. Every downtown street has a story, telling of those people who lived through the Fatimid, Ayyubid, Mamluk, and Ottoman periods, up to the present.

This map highlights some of the most famous historical buildings, well-known characters, and architectural styles, reflecting the mix of treasures that have attracted people to ancient Cairo, and charts more than a thousand years of rich history.

Courtesy of Mai Ghaly Elgazzar.

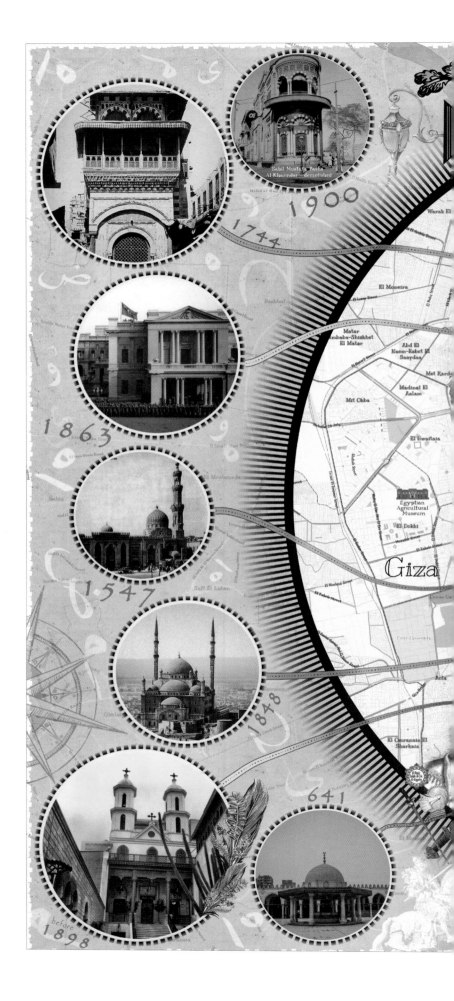

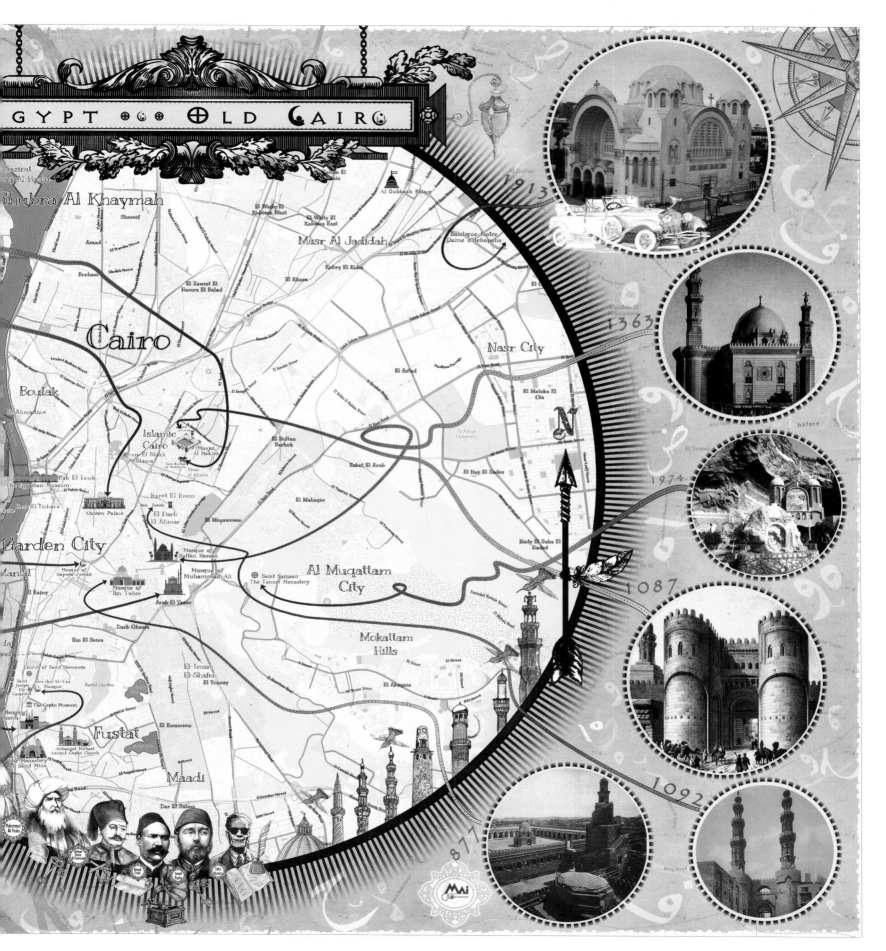

CHANNEL ISLANDS NATIONAL PARK

Tyler Technologies Inc.
Latham, New York, USA
By Seth Frame

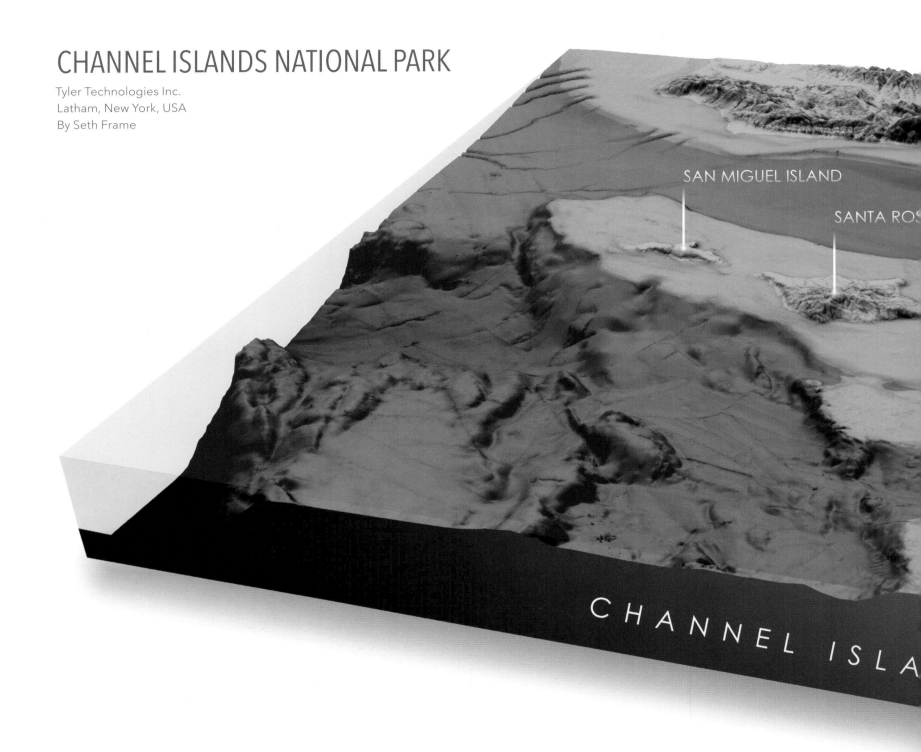

SAN MIGUEL ISLAND

SANTA ROS

CHANNEL ISLA

The Channel Islands consist of eight islands located off the Southern California coast. Five of these islands, shown in the map, constitute the Channel Islands National Park: San Miguel, Santa Rosa, Santa Cruz, Anacapa, and Santa Barbara Islands.

This map, which explores the topography of the islands and the surrounding bathymetric data, shows the changes that fluctuating sea levels may have on the islands.

Courtesy of Tyler Technologies Inc.

SANTA CRUZ ISLAND

ANACAPA ISLAND

SANTA BARBARA ISLAND

S NATIONAL PARK

CONTACT
Seth Frame
Seth.Frame@tylertech.com

SOFTWARE
ArcGIS Pro

DATA SOURCES
ArcGIS Living Atlas of the World

AREAS OF IMPORTANCE FOR THE CONSERVATION OF IMPERILED SPECIES

NatureServe
Arlington, Virginia, USA
By NatureServe

CONTACT
Samantha Lee Belilty
Samantha_Belilty@natureserve.org

SOFTWARE
ArcGIS Pro

DATA SOURCES
NatureServe Network

To better understand where conservation action can help protect at-risk species, NatureServe created this map of biodiversity importance. It is the first-ever high-precision map that predicts locations of imperiled species within the continental US.

Courtesy of NatureServe.

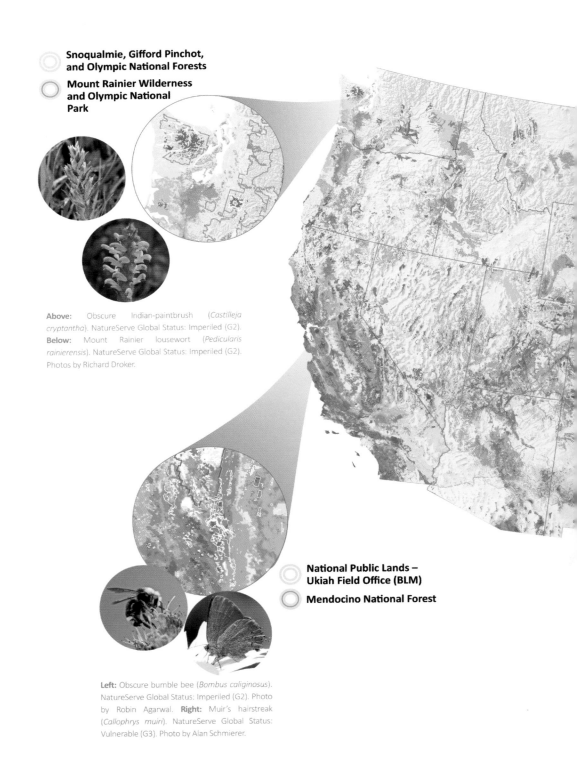

Snoqualmie, Gifford Pinchot, and Olympic National Forests

Mount Rainier Wilderness and Olympic National Park

Above: Obscure Indian-paintbrush (*Castilleja cryptantha*). NatureServe Global Status: Imperiled (G2). **Below:** Mount Rainier lousewort (*Pedicularis rainierensis*). NatureServe Global Status: Imperiled (G2). Photos by Richard Droker.

National Public Lands – Ukiah Field Office (BLM)

Mendocino National Forest

Left: Obscure bumble bee (*Bombus caliginosus*). NatureServe Global Status: Imperiled (G2). Photo by Robin Agarwal. **Right:** Muir's hairstreak (*Callophrys muiri*). NatureServe Global Status: Vulnerable (G3). Photo by Alan Schmierer.

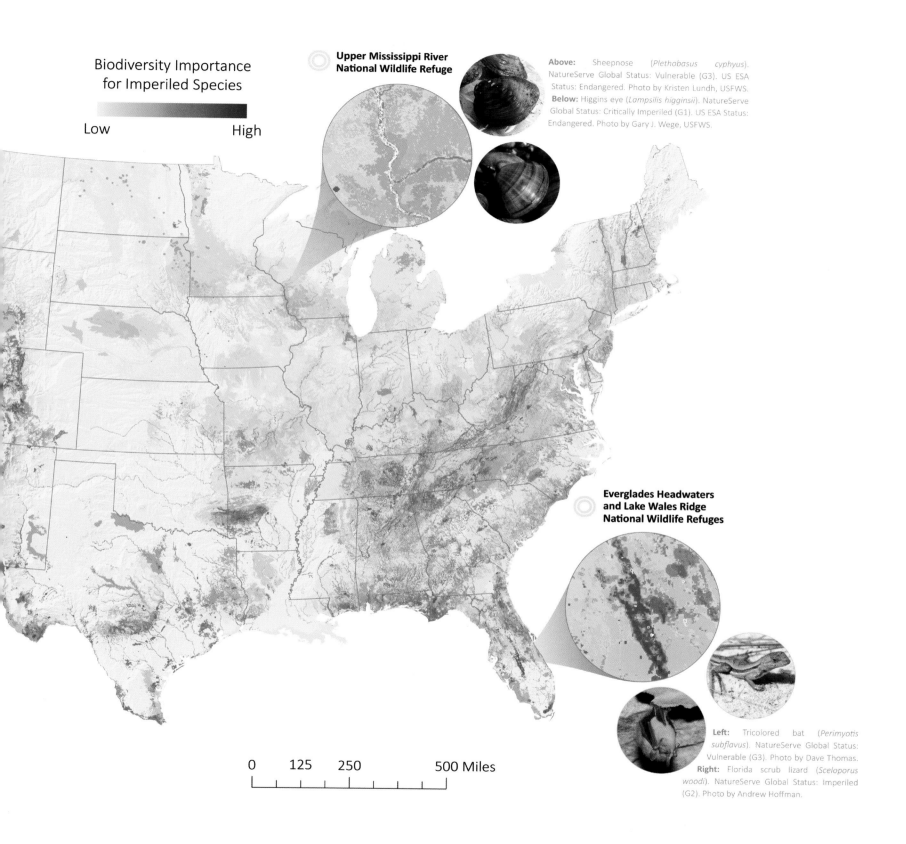

Biodiversity Importance for Imperiled Species

Low   High

Upper Mississippi River National Wildlife Refuge

Above: Sheepnose (*Plethobasus cyphyus*). NatureServe Global Status: Vulnerable (G3). US ESA Status: Endangered. Photo by Kristen Lundh, USFWS. **Below:** Higgins eye (*Lampsilis higginsii*). NatureServe Global Status: Critically Imperiled (G1). US ESA Status: Endangered. Photo by Gary J. Wege, USFWS.

Everglades Headwaters and Lake Wales Ridge National Wildlife Refuges

Left: Tricolored bat (*Perimyotis subflavus*). NatureServe Global Status: Vulnerable (G3). Photo by Dave Thomas. **Right:** Florida scrub lizard (*Sceloporus woodi*). NatureServe Global Status: Imperiled (G2). Photo by Andrew Hoffman.

0   125   250   500 Miles

PROTECTED AREA DATABASE OF THE UNITED STATES (PAD-US)

USGS
Lakewood, Colorado, USA
By GreenInfo Network

CONTACT
Roger Johnson
rjohnson@usgs.gov

SOFTWARE
ArcGIS Pro

DATA SOURCES
State boundaries from Census 2019, TIGER/Line,
US Geological Survey (USGS) Gap Analysis Project (GAP),
2022, Protected Areas Database of the United States
(PAD-US) 3.0: US Geological Survey data release

This map illustrates land and marine areas managed
by government agencies, organizations, and private
land owners. The map is based on the PAD-US version
3.0 published by the US Geological Survey. Lands and
waters included in PAD-US are assigned attributes that
describe management, ownership, and conservation
status measures that denote the level of biodiversity
preservation and other designations that indicate natural,
recreational and cultural uses. The map show US states
only and does not include territories.

Courtesy of USGS.

Manager Name

- Department of Defense
- American Indian Areas (Census Bureau)
- Bureau of Land Management (BLM)
- National Park Service (NPS)
- Forest Service (USFS)
- Army Corps of Engineers (USACE)
- Bureau of Reclamation (USBR)
- U.S. Fish and Wildlife Service (FWS)
- Bureau of Ocean Energy Management (BOEM)
- National Oceanic & Atmospheric Administration (NOAA)
  Marine Sanctuaries & Estuarine Reserves | Other MPAs from NOAA
- Natural Resources Conservation Service (NRCS)
- Other Federal (TVA, ARS, BPA, DOE, etc.)
- State Trust Land
- Other State (NHP, DOT, HS, etc.)
- State Fish and Wildlife
- State Parks and Recreation
- Non-Governmental Organization
- County / Regional Agency Land
- City Land
- Private Conservation; Private Corporation
- Joint, Other, Unknown

Map shows U.S. States only, does not include Territories
This map is based on the PAD-US 3.0 Combined Proclamation (Tribal, DOD only), Marine, Fee, Designation, Easement feature class, published by the USGS Science Analytics and Synthesis (SAS), Gap Analysis Project (GAP).

This map provides a general overview of management, not ownership. Federal and other designated areas may overlap state, private, and other inholdings.

State boundaries from 2019 Census TIGER/Line
More information at: https://usgs.gov/gapanalysis/PAD-US or pad-us@usgs.gov
Map created for USGS by Greeninfo Network, October 2022
Tribal areas included for context.

GEOSPATIAL CONSERVATION AT THE NATURE CONSERVANCY

The Nature Conservancy
Arlington, Virginia, USA
By Zach Ferdaña, Megan Dettenmaier, Chris Bruce, Mrinal Joshi, and Ghufran Zulqisthi

CONTACT
Chris Bruce
cbruce@tnc.org

SOFTWARE
ArcGIS Pro, ArcGIS Maps for Adobe Creative Cloud, Adobe InDesign, Adobe Illustrator

DATA SOURCES
TNC, World Resources Institute,
European Space Agency (ESA),
Shuttle Radar Topography Mission

Geospatial technology systems are playing an ever-growing role in supporting and amplifying The Nature Conservancy's conservation priorities. GIS, remote sensing, and data science are advancing these priorities with ambitious goals for 2030. These goals are grounded in science, supported by geospatial technology, and guided by their mission: to conserve the lands and water on which all life depends.

The Wehea-Kelay landscape, in the province of East Kalimantan, Indonesia, on Borneo, is an area of collaborative management for the protection of orangutan habitat and forest ecosystems. This image, from the cover of our 2021 Geospatial Annual Report & Map Book, shows the Wehea-Kelay boundary and surrounding area, including intact forest landscapes (darker green) and forest cover (lighter green). The Nature Conservancy and partners are working to protect these habitats across Kalimantan by implementing conservation methods that address the commodity-driven land use changes driven by economic growth.

Courtesy of The Nature Conservancy.

WAR AND NATURE

Space Research and Technology Institute,
Bulgarian Academy of Sciences
Bulgaria
By Stefan Stamenov and Vanya Stamenova

CONTACT
Stefan Stamenov
stamenovstefan@yahoo.bg

SOFTWARE
ArcMap

DATA SOURCES
United Nations Institute for Training and Research (UNITAR),
United Nations Satellite Centre (UNOSAT), World Database
on Protected Areas (WDPA), International Union for
Conservation of Nature (IUCN), United Nations Environment
Programme (UNEP)

The protection of conservation areas during war is an issue
that global organizations are grappling with and developing
strategies to address. Detrimental impacts to cultural heritage
sites and infrastructure during armed conflict are well known
and commonly monitored, but areas of environmental
concern have typically been overlooked.

Although the suitability of natural areas for military needs—
such as providing camouflage—makes them more likely to be
affected, their large extents and variety make the detection of
damage more difficult to track. This map presents the location
and proximity of protected areas to conflict zones and
demonstrates a method for detecting the impacts of warfare.

*Courtesy of Dr. Stefan Stamenov and Dr. Vanya Stamenova, Space
Research and Technology Institute, Bulgarian Academy of Sciences.*

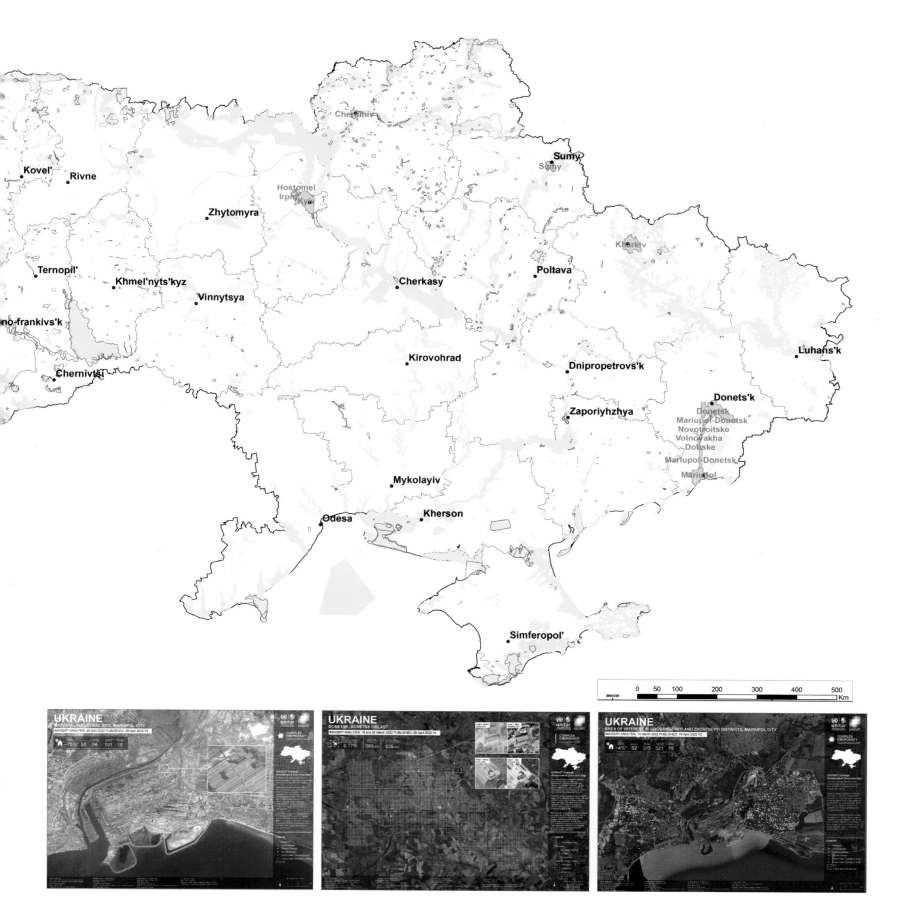

UNTOUCHABLE CARBON

Conservation International
Arlington, Virginia, USA
By Monica Noon, Allie Goldstein, Juan Carlos Ledezma,
Patrick Roehrdanz, Susan Cook-Patton, Seth Spawn-Lee,
T. Max Wright, Mariano Gonzalez-Roglich, David Hole,
Johann Rockström, and Will Turner

CONTACT
Monica Noon
mnoon@conservation.org

SOFTWARE
ArcGIS Desktop, ArcGIS Spatial Analyst™,
Google Earth Engine

DATA SOURCES
Multiple sources cited

Scientists from Conservation International led a team of globally renowned experts that mapped the locations of untouchable carbon—the vast stores of carbon in nature that are vulnerable to release from human activity. If released, this carbon could not be restored by 2050, when the world must reach net-zero emissions to avoid the worst impacts of climate change.

The ecosystems mapped here are the places that humanity cannot afford to destroy. They contain more than 139 billion metric tons of irrecoverable carbon, mostly stored in mangroves, peatlands, old-growth forests, and marshes.

Informed by this pioneering research, Conservation International is undertaking an ambitious initiative to protect more than 1.5 million square miles of ecosystems containing high amounts of irrecoverable carbon and biodiversity.

Courtesy of Conservation International.

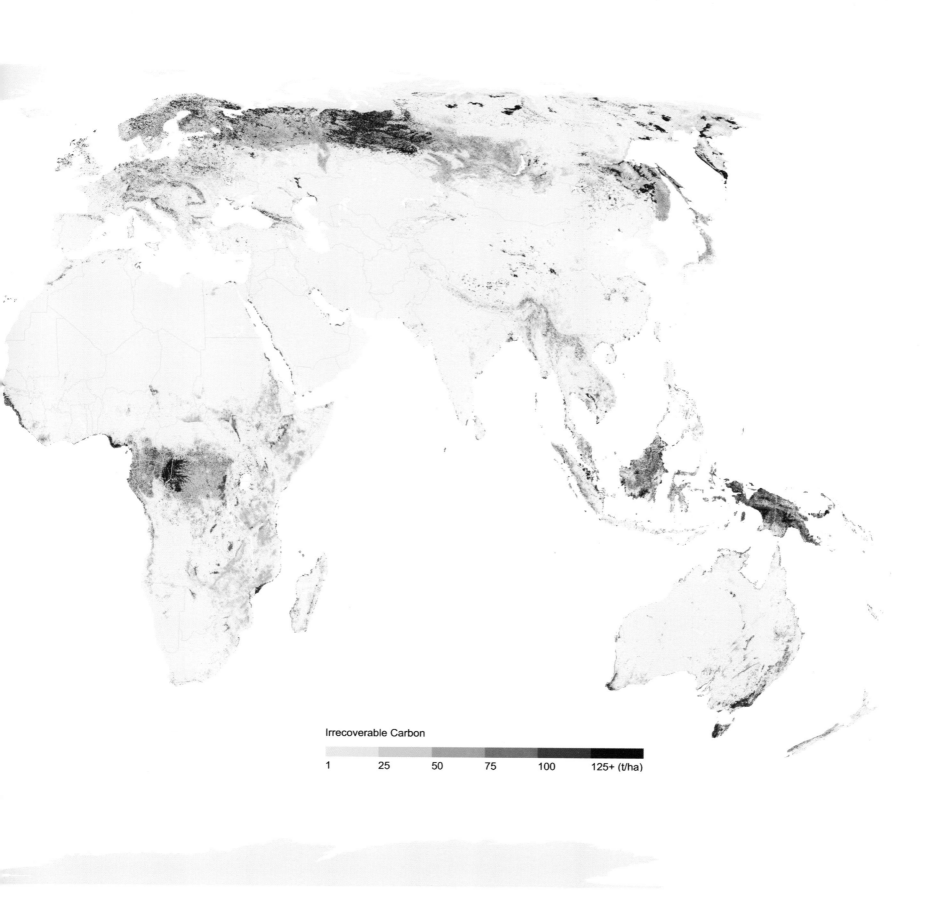

Irrecoverable Carbon

1   25   50   75   100   125+ (t/ha)

AN IMPACT ANALYSIS OF CHINA'S ONE-CHILD POLICY

Tufts University
Somerville, Massachusettes, USA
By Ava Wang

CONTACT
Ava Wang
wxy.ava@gmail.com

SOFTWARE
ArcGIS Pro

DATA SOURCES
China Data Center, National Bureau of Statistics of China

China's former one-child family policy resulted in long-lasting social and cultural implications. To measure its effects, this analysis identified areas in China where the one-child policy was most strictly enforced as of 2000 and evaluated areas in China that witnessed the most significant long-term impacts of this policy as of 2010. Factors used in the analysis included sex ratio, female infant mortality rate, elderly dependency rate, and unmarried male population.

The results of this study show no strong correlation between policy impact and economic development, although faster GDP growth was seen in moderate-impact areas.

Courtesy of Tufts University.

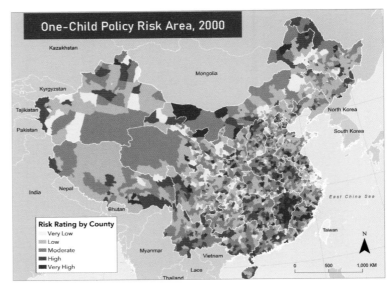

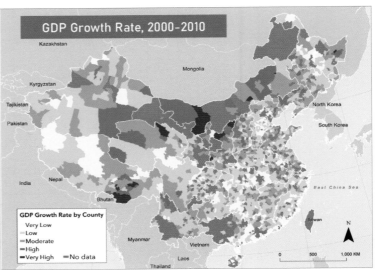

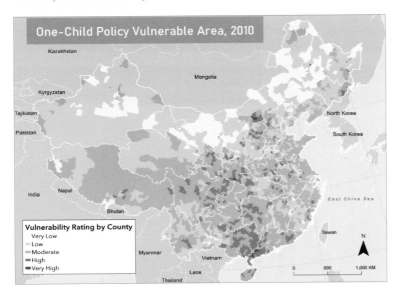

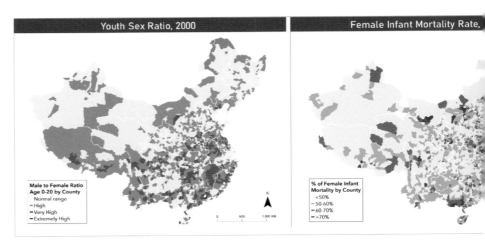

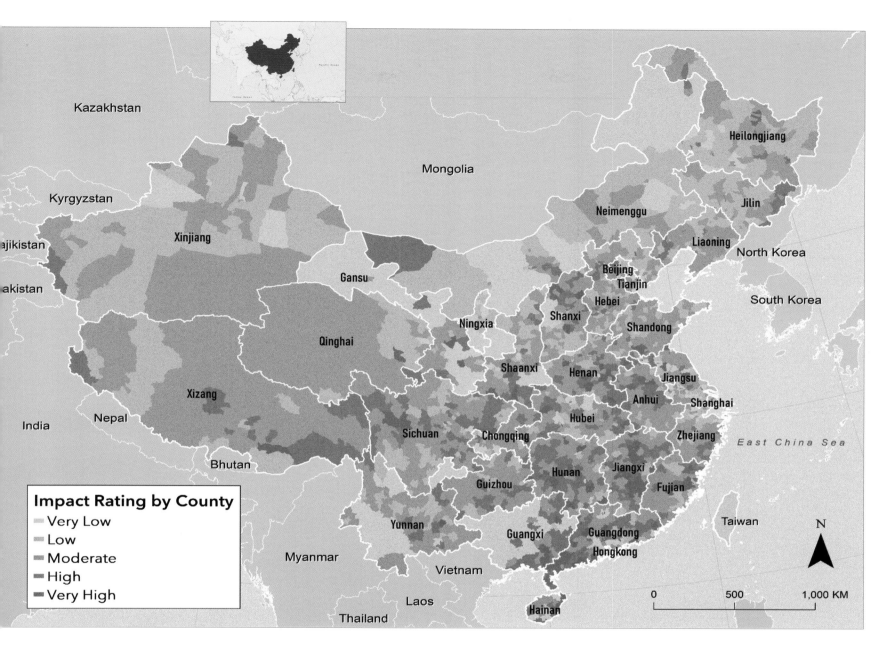

Impact Rating by County

- Very Low
- Low
- Moderate
- High
- Very High

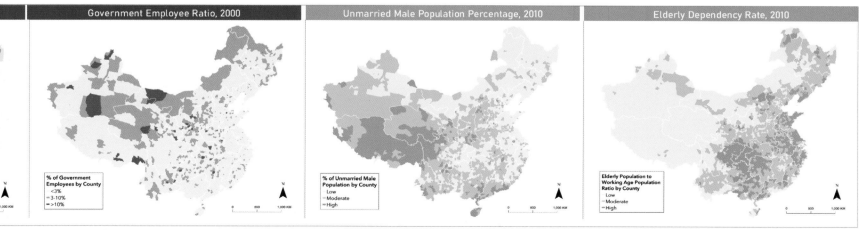

Government Employee Ratio, 2000

% of Government Employees by County
- <3%
- 3-10%
- >10%

Unmarried Male Population Percentage, 2010

% of Unmarried Male Population by County
- Low
- Moderate
- High

Elderly Dependency Rate, 2010

Elderly Population to Working Age Population Ratio by County
- Low
- Moderate
- High

32

DAYTIME POPULATION

Michael Bauer Research GmbH
Nuremberg, Germany
By Michael Bauer Research GmbH

CONTACT
Sebastian Lang
sebastian.lang@mb-research.de

SOFTWARE
ArcGIS Pro

DATA SOURCES
Michael Bauer Research GmbH

The MB-Research Daytime Population Standard indicates how many people are present in a given area on a typical working day (Monday through Friday) between 9:00 a.m. and 6:00 p.m. and assumes that each person stays at a place of main activity during core working hours. This study is based on official statistics of the nighttime population, employed persons by place of home and work, and students by place of home and study.This assessment includes tourism in the calculation.

Courtesy of Michael Bauer Research GmbH.

Daytime Population
- data available
- no data

Madrid - Balance between Day- and Nighttime Population on microgeographical level (percentage)
- below 0 %
- 0 % to 100 %
- 100 % to 200 %
- 200 % to 300 %
- 300 % and above

Hamburg - Balance between Day- and Nighttime Population on postal level (percentage)

below - 30 %   - 30 % to 0 %   0 % to 30 %   30 % to 60 %   60 % and above

Hamburg - Day- and Nighttime Population on postal level

Nighttime Population absolute   Daytime Population absolute

ZAMBIA'S PRESIDENTIAL ELECTION

Bushmaps Cartography
Zambia
By Vincent Abere

CONTACT
Vincent Abere
aberevincent@gmail.com

SOFTWARE
ArcGIS Pro

DATA SOURCES
Electoral Commission of Zambia, Zambia Data Portal

Showing the overwhelming voter turnout during Zambia's 2021 election, this map indicates the winning margins for the top two candidates across the country's 156 constituencies and illustrates the political geography of the country's two main political parties. Presidential candidates are represented with their party colors, and the percentage of total votes won in each constituency is shown.

Courtesy of Vincent Abere, Bushmaps Cartography.

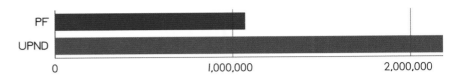

Total Votes Won by Each Party

©Bushmaps

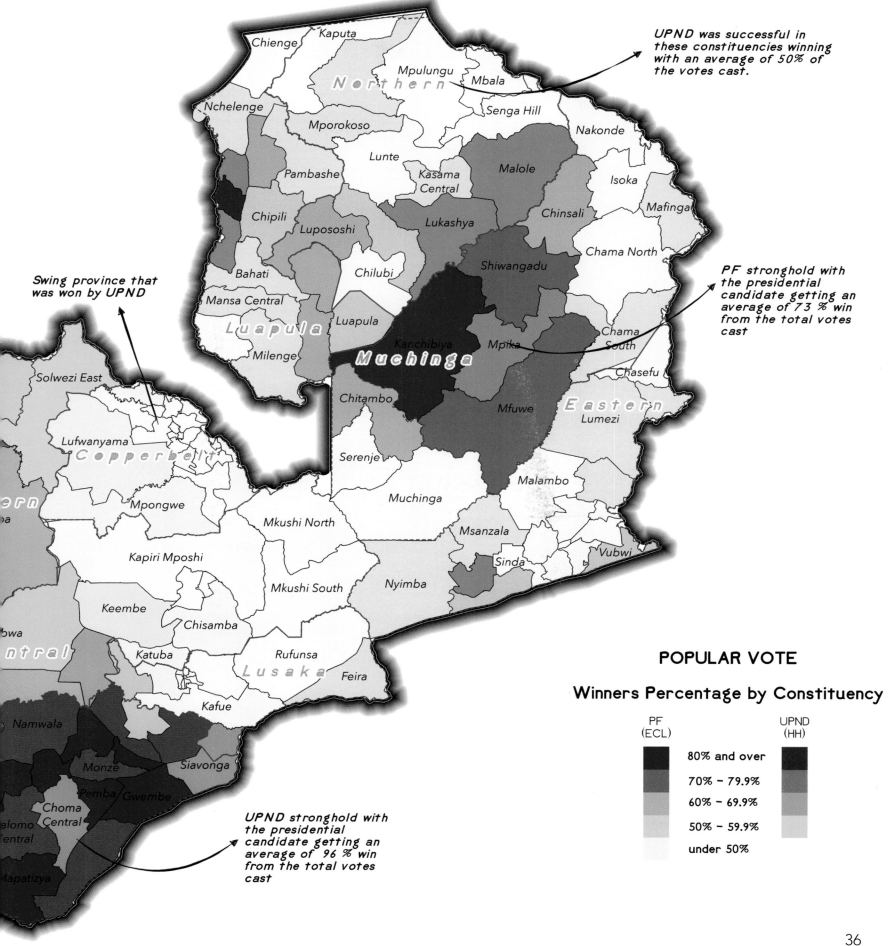

UPND was successful in these constituencies winning with an average of 50% of the votes cast.

Swing province that was won by UPND

PF stronghold with the presidential candidate getting an average of 73 % win from the total votes cast

UPND stronghold with the presidential candidate getting an average of 96 % win from the total votes cast

POPULAR VOTE

Winners Percentage by Constituency

PF (ECL)		UPND (HH)
	80% and over	
	70% – 79.9%	
	60% – 69.9%	
	50% – 59.9%	
	under 50%	

THE GENDER GAP IN STEM

Johns Hopkins University
Baltimore, Maryland, USA
By Emily Long

CONTACT
Emily Long
jlong44@jh.edu

SOFTWARE
ArcGIS Pro, Adobe Illustrator

DATA SOURCES
US Census Bureau (2020)

Women are historically underrepresented in science, technology, engineering, and math (STEM) fields and are also notably paid less than men. Using symbols and colors, this map represents two datasets—the percentage of women in STEM careers by state in the US and the wage gap between women and men.

Courtesy of Johns Hopkins University.

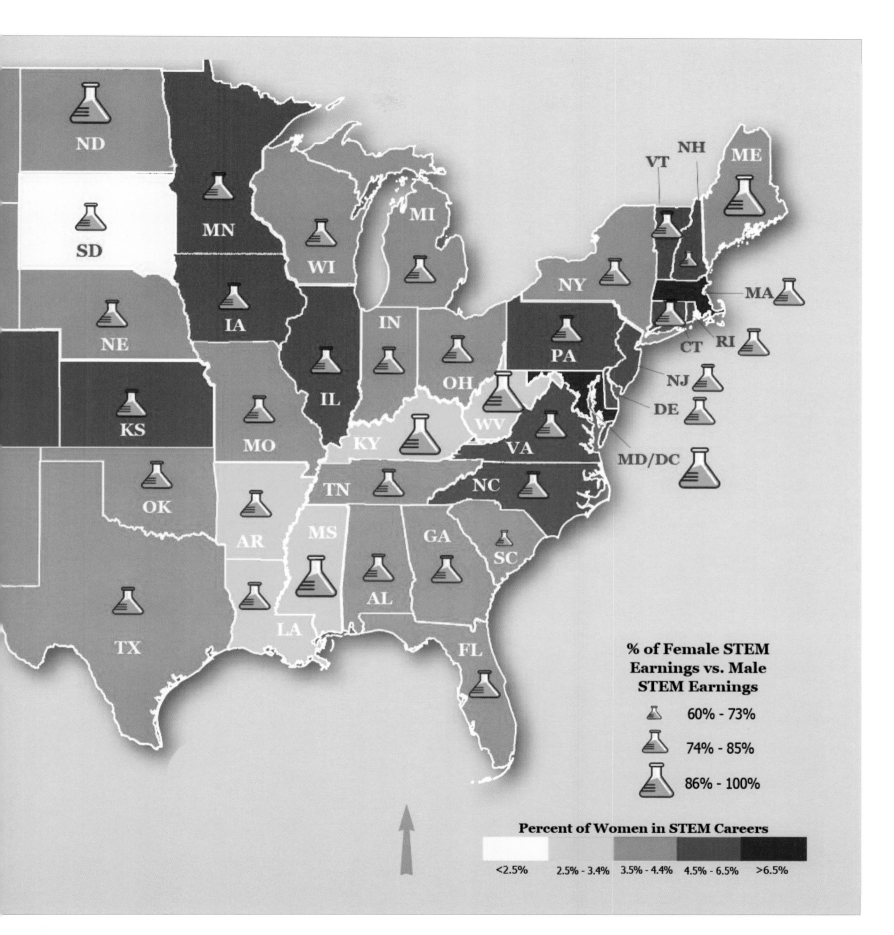

% of Female STEM
Earnings vs. Male
STEM Earnings

60% - 73%

74% - 85%

86% - 100%

Percent of Women in STEM Careers

<2.5%   2.5% - 3.4%   3.5% - 4.4%   4.5% - 6.5%   >6.5%

SOLAR ENERGY POTENTIAL IN CANADA

ABO Wind Canada Ltd.
Calgary, Canada
By Dave Taylor

CONTACT
Dave Taylor
daveataylor1@gmail.com

SOFTWARE
ArcGIS Pro, ArcGIS Spatial Analyst, QGIS, Microsoft Excel

DATA SOURCES
Canadian Weather Energy and Engineering Datasets (CWEEDS),
Natural Resources Canada (NRCan), Airbus, USGS,
National Geospatial-Intelligence Agency (NGA),
National Aeronautics and Space Administration (NASA), CGIAR,
National Center for Ecological Analysis and Synthesis (NCEAS),
National Longitudinal Surveys (NLS), OS, NMA,
Geodatastyrelsen, US General Services Administration (GSA),
GeoSpatial Innovations (GSI) and the GIS User Community,
Esri, Garmin, Food and Agriculture Organization (FAO),
National Oceanic and Atmospheric Administration (NOAA)

Finding areas with high solar energy potential is an important
part of choosing locations for new solar energy projects.
In this study, solar data was collected from weather monitoring
stations throughout Canada. It was converted, processed,
merged, and summarized. A final product was created that
filled in the missing values between weather stations and gives
a general understanding of solar energy potential across the
country. This overview shows the annual potential solar energy by
direct normal irradiance and provides a valuable tool for future
energy planning.

Courtesy of ABO Wind Canada Ltd.

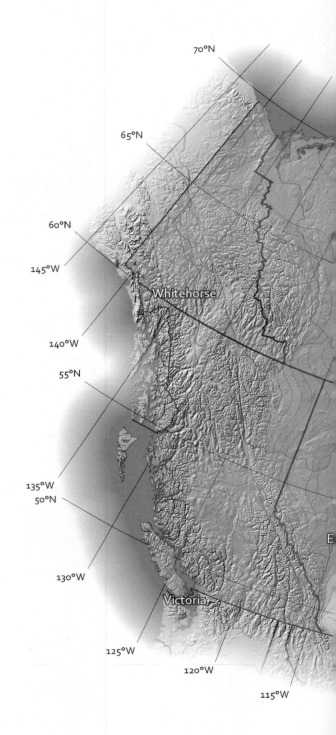

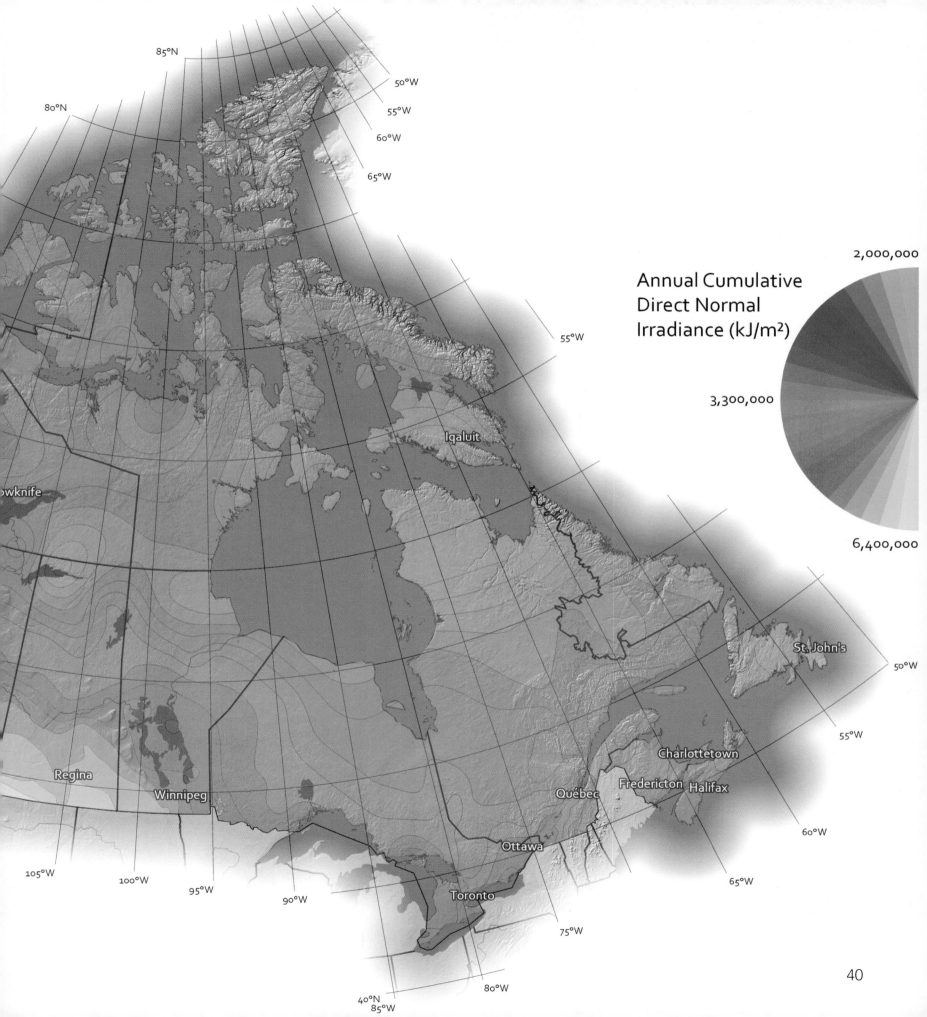

Annual Cumulative
Direct Normal
Irradiance (kJ/m²)

2,000,000

3,300,000

6,400,000

Iqaluit

wknife

St. John's

Regina

Winnipeg

Charlottetown

Québec

Fredericton Halifax

Ottawa

Toronto

85°N

80°N

50°W

55°W

60°W

65°W

55°W

50°W

55°W

60°W

65°W

105°W

100°W

95°W

90°W

75°W

80°W

40°N

85°W

THE GEOLOGY OF MIAMI COUNTY, KANSAS

Kansas Geological Survey, University of Kansas
Lawrence, Kansas, USA
By John W. Dunham, Anthony L. Layzell, K. David Newell,
Stephan C. Oborny, and Rolfe D. Mandel

CONTACT
John Dunham
jdunham@ku.edu

SOFTWARE
ArcMap, ArcGIS Spatial Analyst, Profile Tool

DATA SOURCES
Layzell, Newell, Oborny, Mandel's field mapping

Miami County, in eastern Kansas, is an area of east-facing ridges alternating with flat to gently rolling plains. This geology map shows bedrock and sediment layers either at the surface or immediately under the soil and is the first detailed geologic mapping of the area since 1966.

The youngest deposits shown on the map are found in valleys and were deposited within the last 11,000 years. Most surface rocks formed from sediment were deposited in ancient seas about 290–310 million years ago. The underlying strata are Pennsylvanian limestone and shale that dip gently to the west and northwest.

Courtesy of Kansas Geological Survey, University of Kansas.

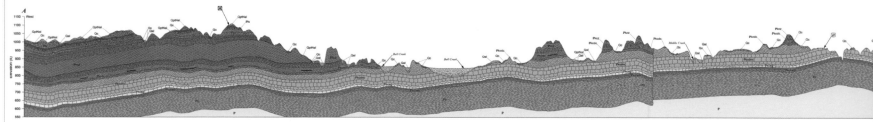

MORE THAN 3,600 YEARS OF TSUNAMI HISTORY

NOAA
Boulder, Colorado, USA
By Jesse Varner and Nicolás Arcos

CONTACT
Jesse Varner
Jesse.Varner@noaa.gov

SOFTWARE
ArcGIS Pro, Adobe Illustrator

DATA SOURCES
National Centers for Environmental Information/World Data Service Global Historical Tsunami Database, Natural Earth, Plate Boundaries: P. Bird, 2003

NOAA's National Centers for Environmental Information (NCEI) and World Data Service (WDS) for Geophysics, together with the International Tsunami Information Center (ITIC), collaborated to produce this map showing the causes of historical tsunamis from 1610 BCE to 2022 CE.

NCEI's Global Historical Tsunami, Significant Earthquake, and Significant Volcanic Eruption databases provide valuable information that agencies use to issue alerts when potentially deadly or damaging events occur. These databases are publicly available and play a critical role in informing the public about natural hazards.

Tsunami sources on the map are symbolized by cause (earthquake, volcanic eruption, landslide, or unknown) and are color-coded by the number of deaths caused by the tsunami. The 10 deadliest tsunamis and those that were triggered by a magnitude 9 (or greater) earthquake are labeled with the event date. For context on seismic activity, the plate boundaries and approximate velocity at each plate interface are shown.

Courtesy of the US Department of Commerce NOAA.

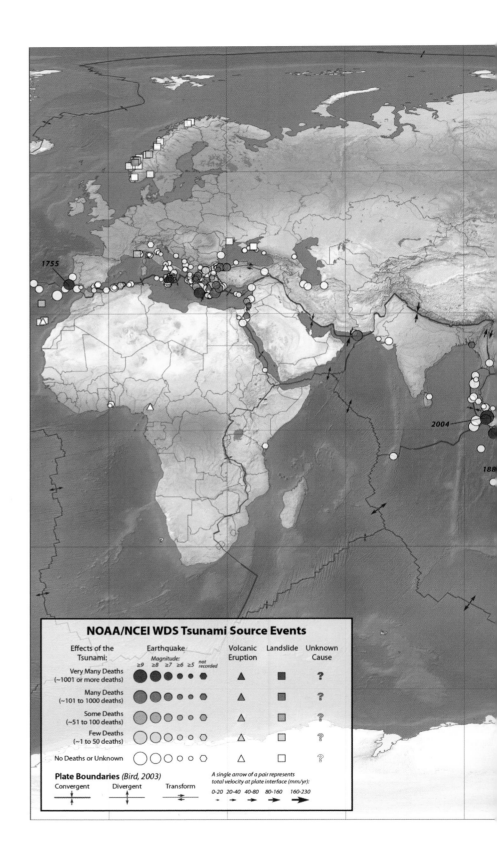

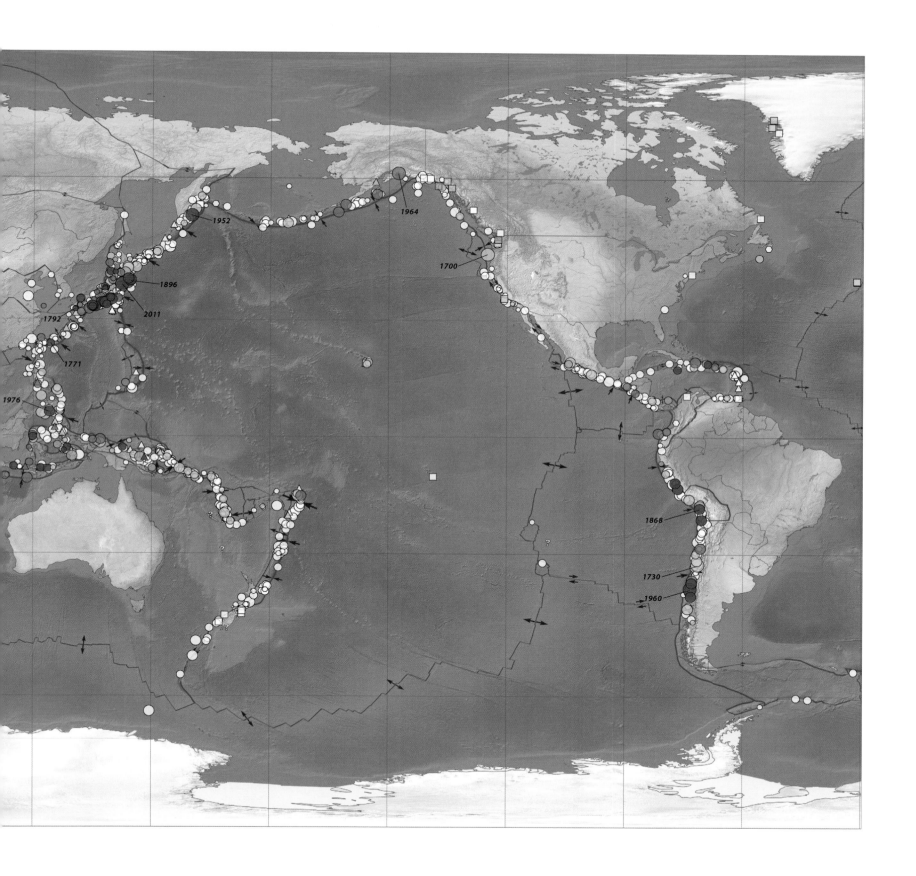

GEOTHERMAL ENERGY OPPORTUNITIES IN SOUTH AMERICA

Getech Group PLC
London, United Kingdom
By Chris Jepps

CONTACT
Getech
info@getech.com

SOFTWARE
ArcGIS Image Server, ArcGIS Online

DATA SOURCES
Getech Group PLC

The Heat Seeker web app was developed to locate potential sites for accessing geothermal energy. It uses advanced spatial analysis and machine learning algorithms to integrate geophysical, geologic, commercial, and public data to create maps of geothermal suitability. This image shows an analysis of geothermal opportunity in South America and highlights in green where geothermal energy projects may be preferable to wind or solar development.

Courtesy of Getech Group PLC.

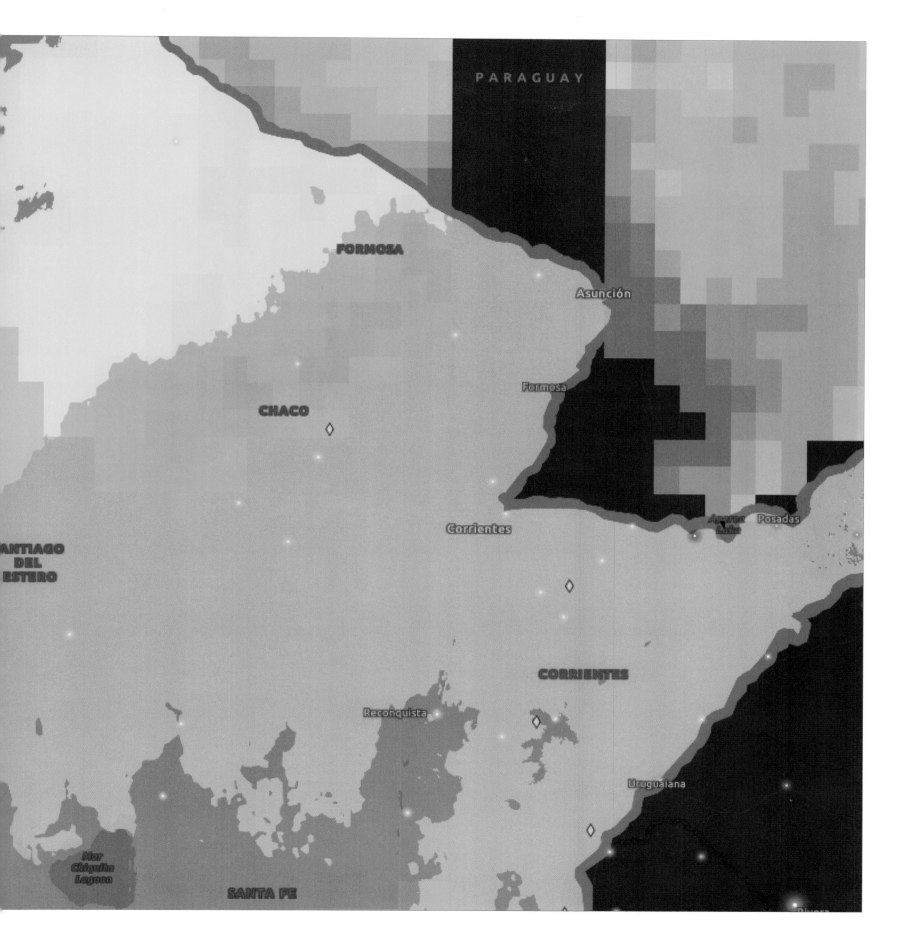

UNDERGROUND OIL AND GAS CAVERNS AT A STORAGE FACILITY IN GERMANY

Research Center of Post-Mining, Bochum, Germany,
By Benjamin Haske

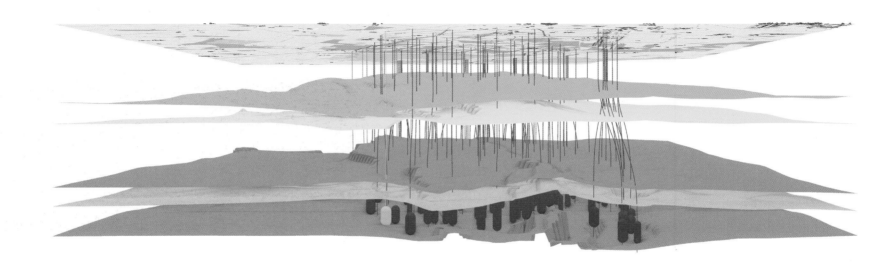

These side and transparent views show underground caverns at one of the largest oil and gas storage facilities in Europe. Liquid and gas energy sources are temporarily stored under high pressure in more than 100 gigantic salt caverns at a depth of more than 1,000 meters. Each color represents a different material: green equals natural gas, red equals mineral oil, yellow equals helium, blue equals brine. The colors of the geologic layers follow international standards and show that the caverns are in a thick salt horizon.

The map was created as part of an integrated safety monitoring system being developed as part of a research project. Based on the highly accurate locations of caverns, pipelines, and land use on the surface, the areas to be monitored can be easily identified.

Courtesy of Research Center of Post-Mining

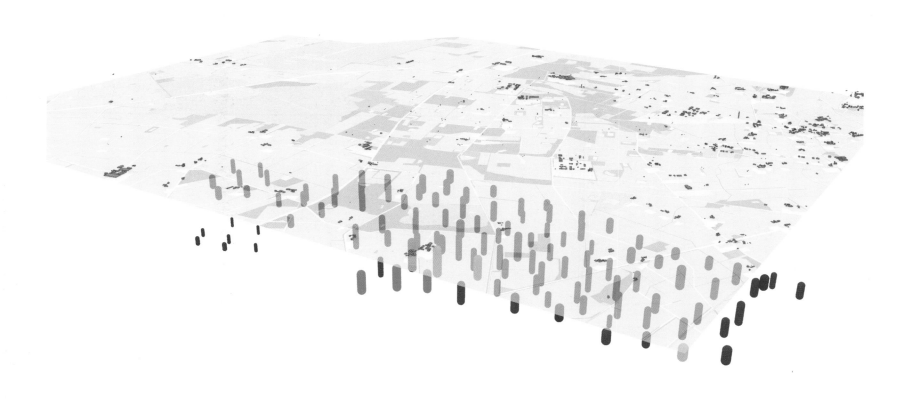

CONTACT
Benjamin Haske
benjamin.haske@thga.de

SOFTWARE
ArcGIS Pro, ArcGIS 3D
Analyst™

DATA SOURCES
Research Center of Post-Mining at the Technische Hochschule
Georg Agricola University; GEOportal NRW (Bezirksregierung Köln);
Salzgewinnungsgesellschaft Westfalen mbH;
Uniper Energy Storage GmbH

WILDFIRE RISK MANAGEMENT FOR SMALL ELECTRIC UTILITIES

Clatskanie People's Utility District
Clatskanie, Oregon, USA
By Sephe Fox

CONTACT

Sephe Fox
sfox@clatskaniepud.com

SOFTWARE

ArcMap, ArcGIS Spatial Analyst, ModelBuilder™

DATA SOURCES

Oregon Lidar Consortium, LANDFIRE (2016 remap),
SILVIS (2010), OpenStreetMap, Clatskanie PUD GIS/OMS

Catastrophic wildfire is an increasing problem in the northwestern US. Mitigating the inherent risk of a utility-sparked fire can be expensive, so identifying locations with higher risk can allow mitigation efforts to be more effective.

For a small utility with limited resources, the traditional options for wildfire risk analysis can be unsatisfactory. Hiring GIS consultants can be expensive. A manual analysis can assign general risk scores to regions but may generate inaccurate or biased results. Using analyses from government agencies can be inadequate because the products often rely on coarse resolution and lack a utility perspective.

Clatskanie People's Utility District decided to conduct its own analysis using publicly available and utility-specific data to model risk factors for fire ignition, rapid spread, and catastrophic impacts. The analysis identified higher fire risk areas where mitigation can reduce the likelihood of a utility-caused fire—increasing public safety while minimizing operating costs and keeping utility rates low.

Courtesy of Clatskanie People's Utility District.

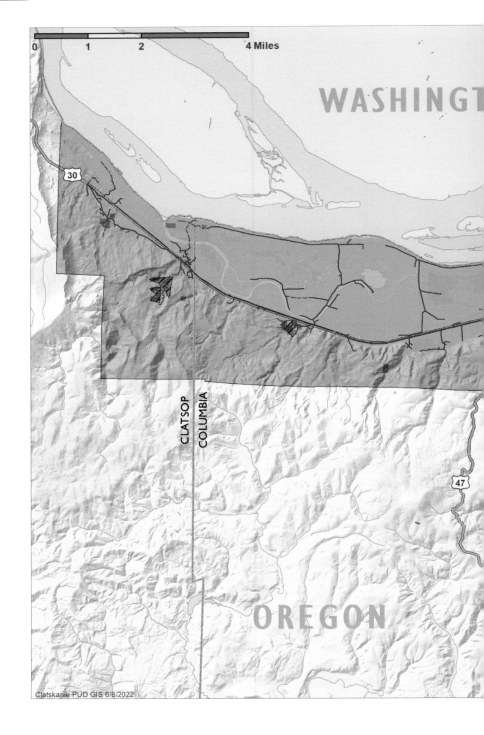

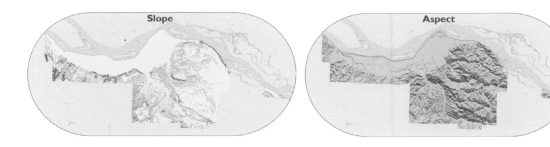

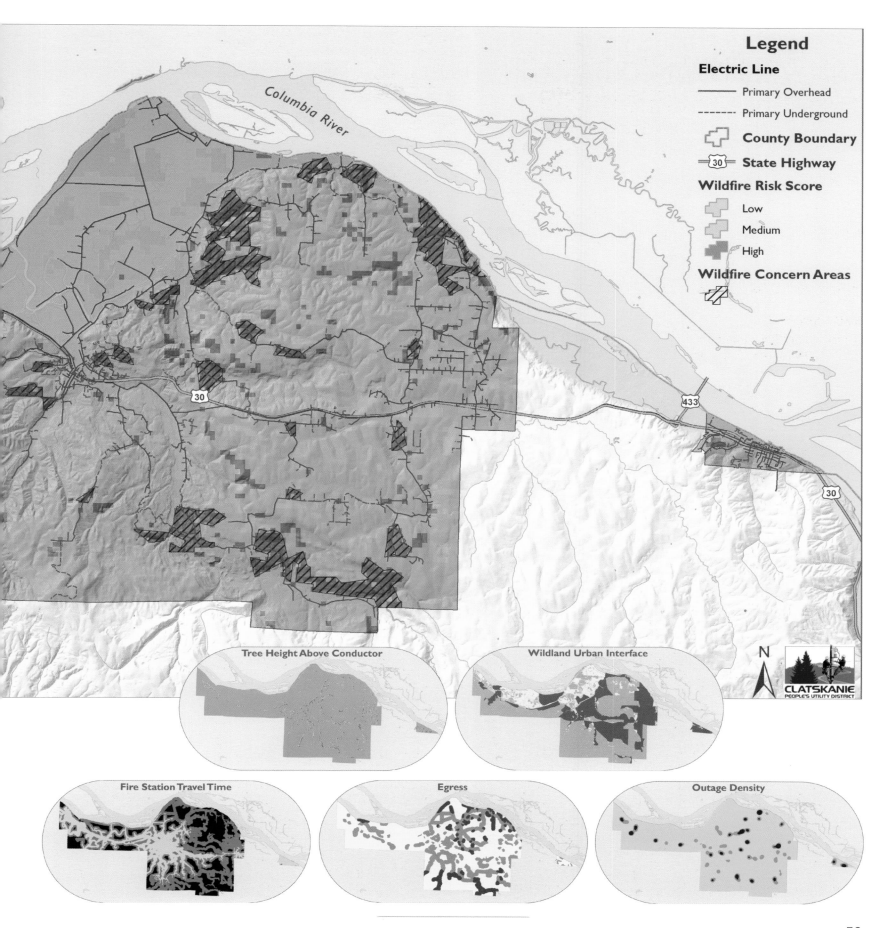

Legend

Electric Line

——— Primary Overhead

------- Primary Underground

County Boundary

30 **State Highway**

Wildfire Risk Score

Low

Medium

High

Wildfire Concern Areas

Columbia River

Tree Height Above Conductor

Wildland Urban Interface

Fire Station Travel Time

Egress

Outage Density

N

CLATSKANIE
PEOPLE'S UTILITY DISTRICT

ANALYZING POWER OUTAGES IN 3D

Grand Valley Rural Power, Fruita, Colorado, USA
By Ethan Schaecher

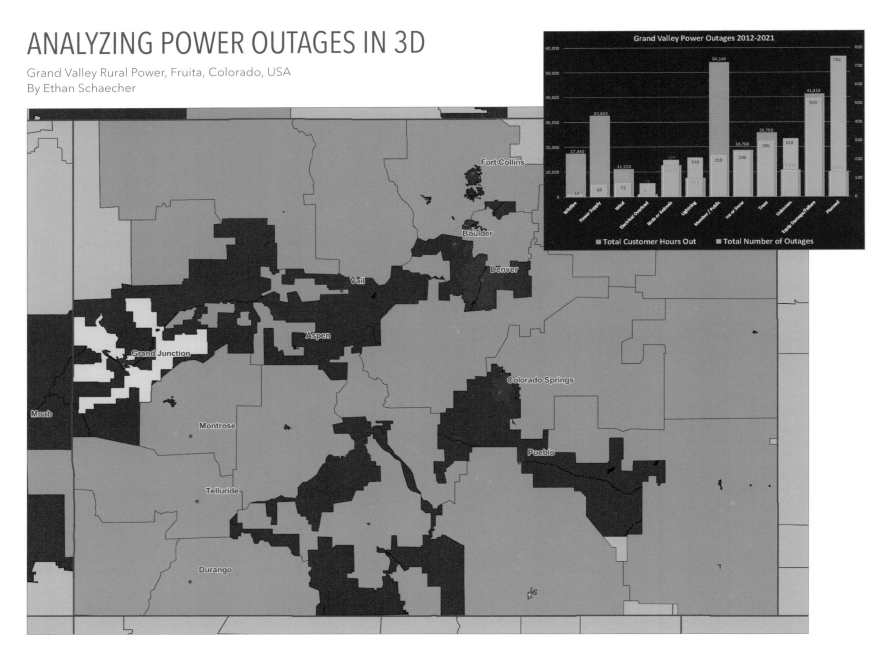

From 2012 to 2021, Grand Valley Power experienced nearly 2,700 power outages throughout its electrical cooperative service territory in western Colorado. To analyze these outages, all outage records were assigned a location and categorized into 12 different causes. Points for each outage were aggregated into hexagons and extruded in 3D based on the number of customers affected multiplied by the duration of the outage in minutes.

This research demonstrates that spatial analysis is an important tool in mitigating future outages by accomplishing two major objectives: saving the cooperative money and increasing reliability for more than 18,000 customers.

Courtesy of Grand Valley Rural Power.

CONTACT

Ethan Schaecher
ethan.d.schaecher@gmail.com

SOFTWARE

ArcGIS Pro, ArcGIS Online, ArcGIS StoryMaps, National Information Solutions Cooperative (NISC) Outage Management System (OMS), Notepad++, Microsoft Excel

DATA SOURCES

Grand Valley Power Outage History Tabular Records—National Information Solutions Cooperative (NISC), Outage Management System (OMS), Esri basemaps, Cities & Towns: US Census

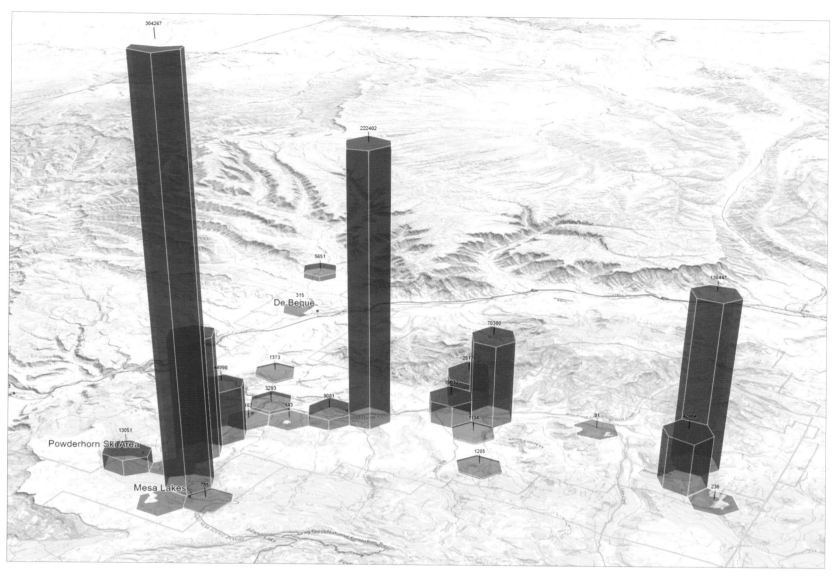

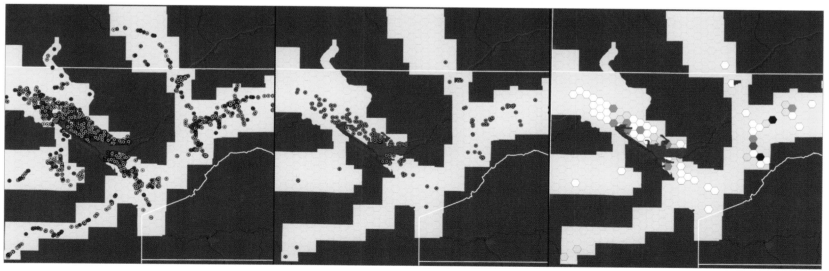

TRACKING THE DISPORPORTIONATE IMPACTS OF TOXIC AIR

University of Kentucky,
Lexington, Kentucky, USA
By Sarah Roach

This series of maps explores the connection between polluting facilities—identified through the EPA's Toxic Release Inventory (TRI)—and the health of marginalized residents in Louisville, Kentucky.

By using bivariate symbology, these maps show when different factors are present. The map identifying census tracts with high rates of asthma and chronic obstructive pulmonary disease (COPD) shades the locations in purple if both rates are high. The map identifying census tracts with a high percentage of Black residents and high rates of poverty also uses purple to indicate areas where both occur. Most TRI facilities are in the historically Black west-end and ethnically diverse south-end tracts, areas with high poverty rates and high to medium rates of both COPD and asthma.

Courtesy of University of Kentucky.

Toxics Release Inventory (TRI) Facilities

RSEI Score

- ∘   0 – 138451
- ○   138452 – 534916
- ○   534917 – 871959
- ○   871960 – 5351150
- ○   5351151 – 11306718

Percent of Black Residents and Percent Below Poverty

Black Residents: 0.3-8.6%   8.6-20.6%   20.6-98.9%

Percent Below Poverty: 0.9-7.6%   7.6-18.9%   18.9-81%

0   4   8   15 Miles

CONTACT
Sarah Roach
sarahroach20@gmail.com

SOFTWARE
ArcGIS Pro

DATA SOURCES
Environmental Protection Agency (EPA), Centers for Disease Control and Prevention (CDC) PLACES, 500 Cities Project, Census Bureau, American Community Survey (ACS) 2019 5-Year Estimates, EPA

TRACKING MOSQUITO OUTBREAKS AND DENGUE FEVER

Município de Itajaí
Itajaí, Santa Catarina, Brazil
By Lucas Pereira Cabral and Murilo Sodré

CONTACT
Lucas Pereira Cabral
setec@itajai.sc.gov.br

SOFTWARE
ArcGIS Pro, Experience Builder, ArcGIS Dashboards

DATA SOURCES
Municipal Health Department of Itajaí (SMS), Municipal
Department of Technology of Itajaí (SETEC)

This dashboard tracks efforts to control the *Aedes aegypti*
mosquito, which can spread dengue fever and other diseases.
Key indicators in this fight are easy to see and provide near-
real-time metrics of the on-site work performed by technicians
in the municipality of Itajaí, Brazil. This tool allows public
officials to access updated details, including the location
of positive cases, traps, and visits. The density of mosquito
outbreaks is illustrated using a heat map.

Courtesy of Município de Itajaí.

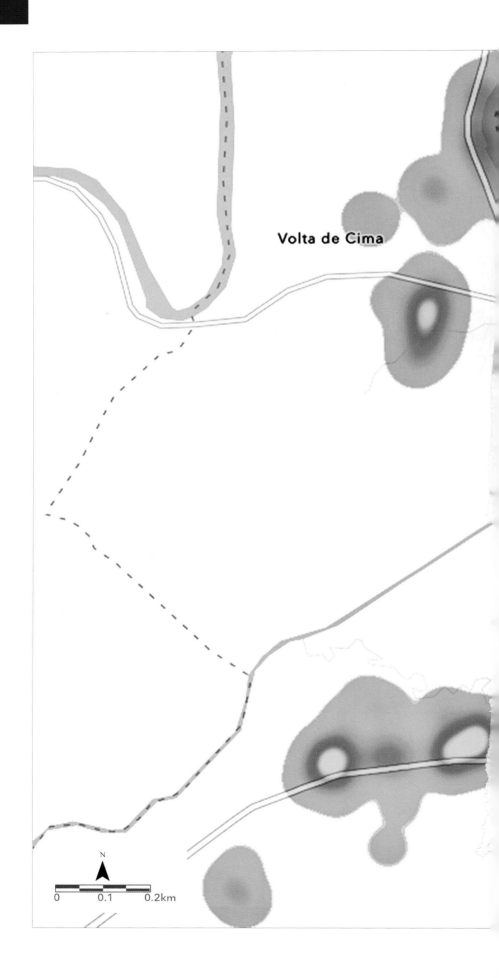

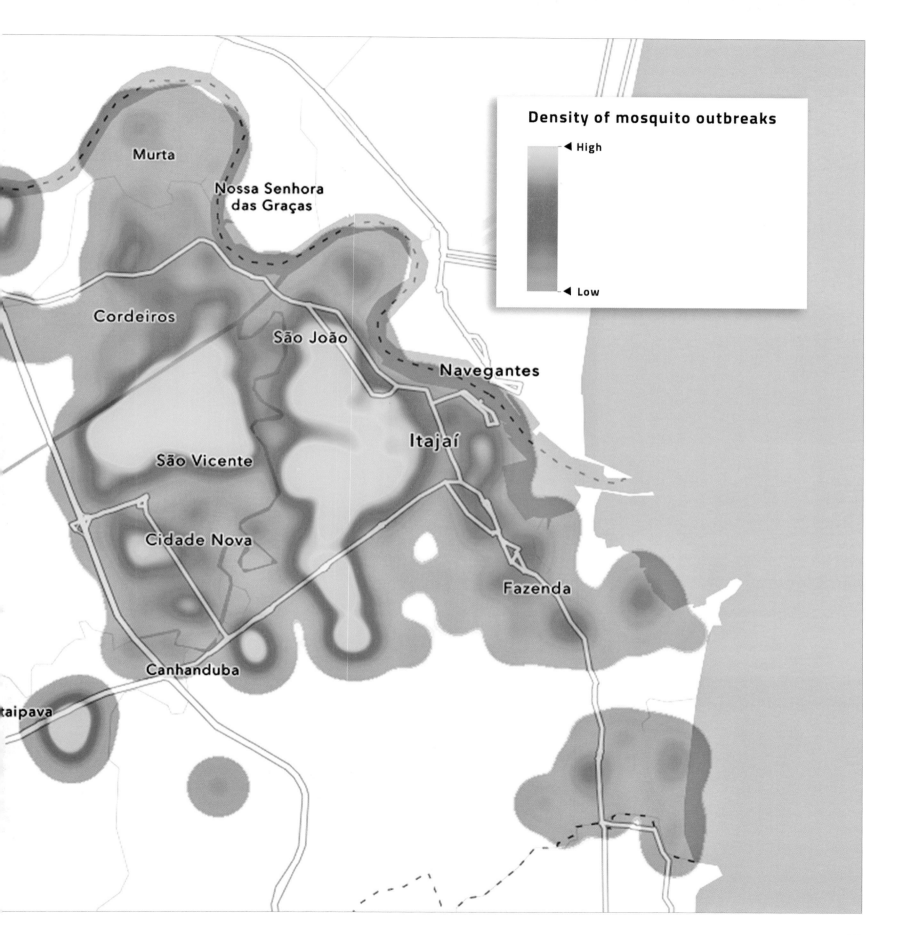

Density of mosquito outbreaks

◄ High

◄ Low

Murta

Nossa Senhora
das Graças

Cordeiros

São João

Navegantes

São Vicente

Itajaí

Cidade Nova

Fazenda

Canhanduba

taipava

ARCTIC AND GREAT LAKES ICE MAPPING

US National Ice Center, Suitland, Maryland, USA
By Mark Denil

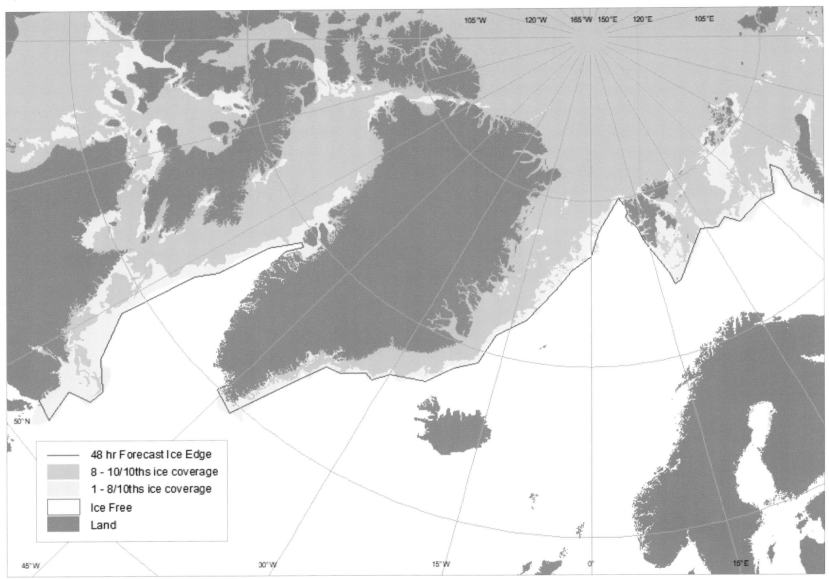

Ice thickness and extent change constantly, creating a highly dynamic environment. For this reason, the US National Ice Center (USNIC) monitors polar ice conditions on a daily and weekly basis—providing public information on ice conditions that can be used by vessels sailing near, in, or under icy waters. The Arctic daily forecast describes current conditions and provides a 48-hour forecast of both the Atlantic and Pacific sides of the Arctic. USNIC ice analysts and forecasters oversee the daily creation of these maps, which are automatically generated using scripts. Symbology focuses attention on the forecast line, and the scale and extents of each map are set automatically.

The US Coast Guard District 9 Ice Chart is a four-panel briefing map that is prepared twice weekly for strategic and tactical briefings at Coast Guard headquarters. This map brings together detailed information on Great Lakes ice thickness and concentration to summarize current and recent conditions. This supports operational decision-making for resource allocation and planning.

Courtesy of US National Ice Center.

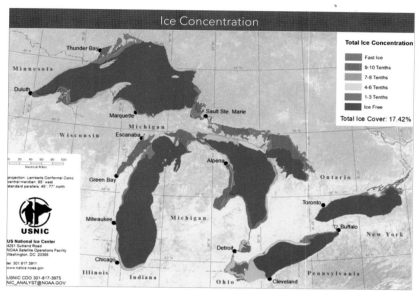

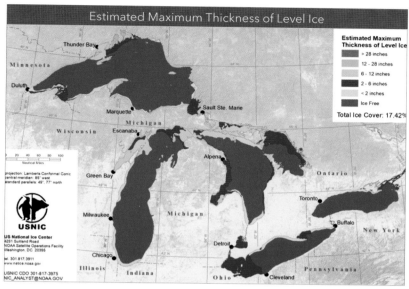

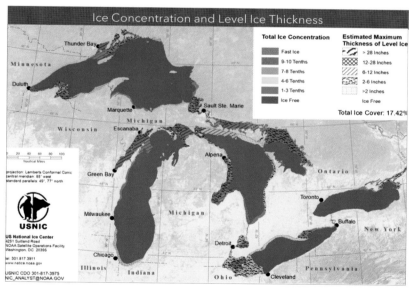

CONTACT
USNIC Ice Desk
nic_analyst@noaa.gov

SOFTWARE
ArcMap™

DATA SOURCES
US National Ice Center (USNIC)

STUDYING THE WOLVES OF DENALI NATIONAL PARK

Johns Hopkins University
Traverse City, Michigan, USA
By Christopher Johnson

CONTACT
Christopher Johnson
cjohn323@jhu.edu

SOFTWARE
ArcGIS Pro

DATA SOURCES
USGS, NPS

Denali National Park's wolf packs are tracked by the National Park Service (NPS) and are shown here with their territorial ranges. Each wolf pack symbol is proportional to the current size of the pack and indicates that pack's recent growth or loss. The Denali wolf population grew from 68 wolves in March 2020 to 93 wolves in March 2021—nearly 27 percent. New wolf packs are generally appearing in regions of higher elevation, possibly a sign of the wolves' adaptation to the melting permafrost.

Courtesy of Johns Hopkins University.

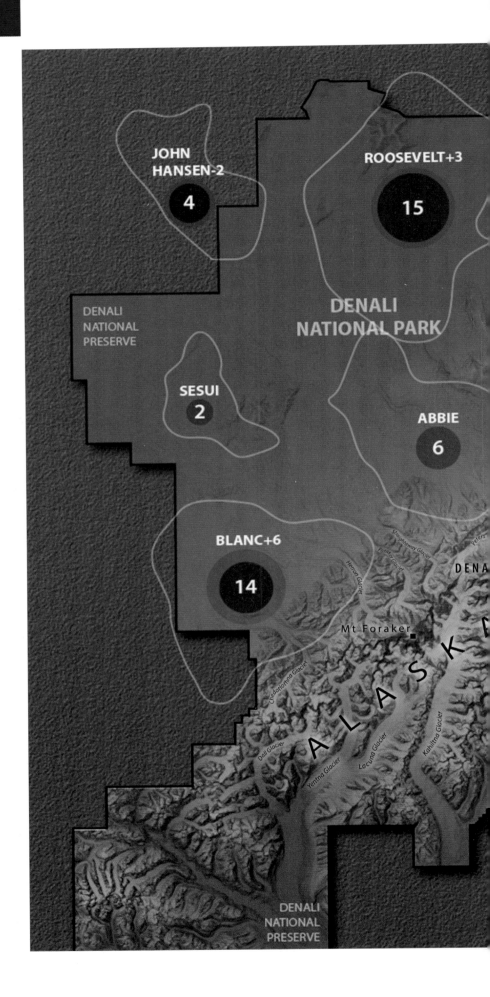

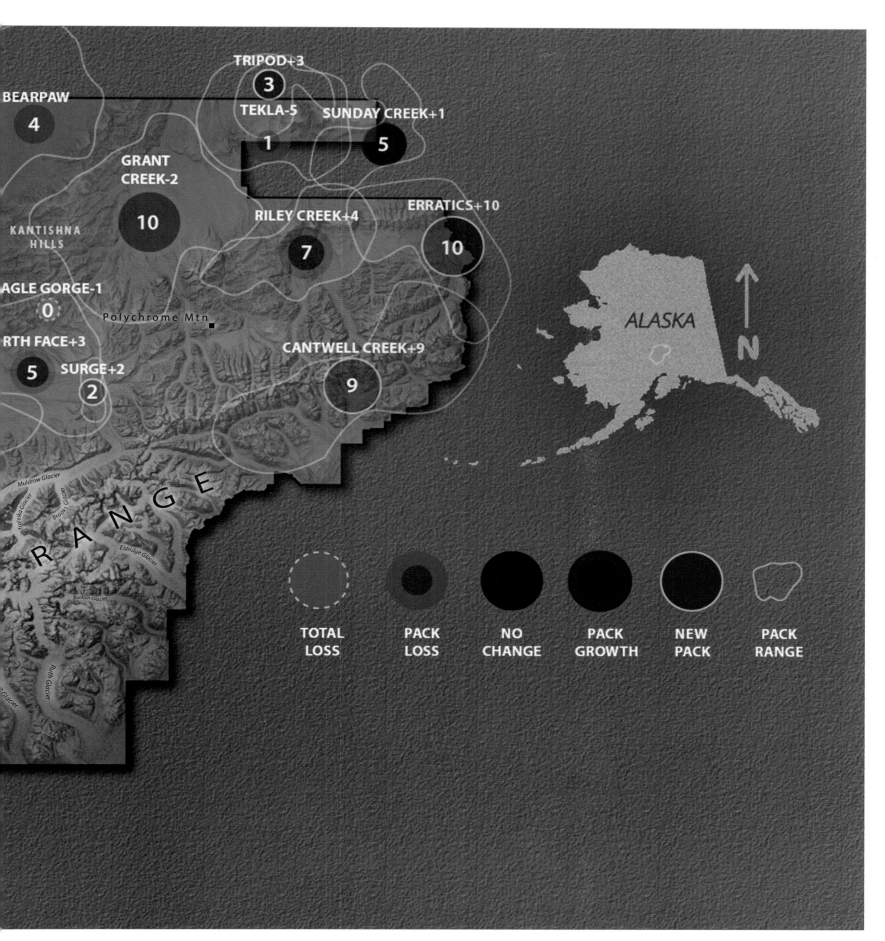

BEARPAW
4

TRIPOD+3
3

TEKLA-5
1

SUNDAY CREEK+1
5

GRANT
CREEK-2
10

KANTISHNA
HILLS

RILEY CREEK+4
7

ERRATICS+10
10

AGLE GORGE-1
0

Polychrome Mtn

RTH FACE+3
5

SURGE+2
2

CANTWELL CREEK+9
9

ALASKA

N

Muldrow Glacier

R A N G E

Brooks Glacier

Traleika Glacier

Eldridge Glacier

Buckskin Glacier

Ruth Glacier

Glacier

TOTAL
LOSS

PACK
LOSS

NO
CHANGE

PACK
GROWTH

NEW
PACK

PACK
RANGE

TRACKING THE CHANGING VEGETATION IN ABU DHABI

iSpatial Techno Solutions, Abu Dhabi, United Arab Emirates
By Narendra Vattem

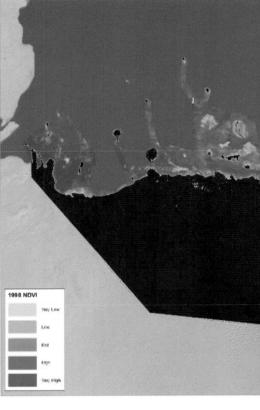

The map shows the changes in vegetation inside the Emirate of Abu Dhabi from 1988 to 2016. Using normalized difference vegetation index (NDVI) analysis, it documents natural and planted vegetation—including agriculture—across urban and rural areas and shows changes in each municipality.

Using geostatistical models to quantify and qualify vegetation, the map shows the presence or absence of plants and indicates their relative health. This analysis supports conservation and pollution reduction and provides benchmarks that can be referenced in future analysis of environmental change.

Courtesy of iSpatial Techno Solutions.

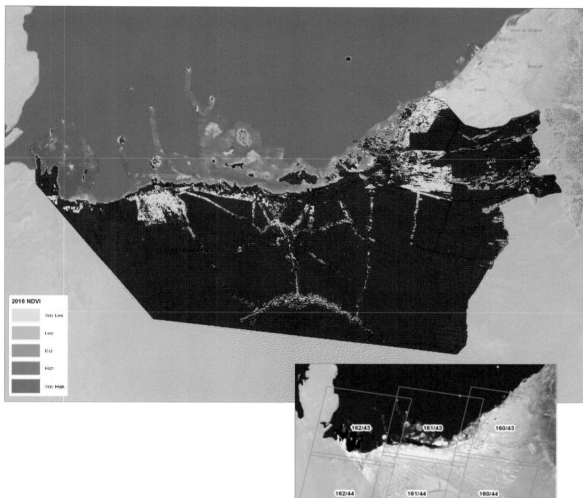

2016 NDVI
- Very Low
- Low
- Mid
- High
- Very High

162/43 161/43 160/43

162/44 161/44 160/44

CONTACT
Narendra Vattem
narendra.vattem@ispatialtec.com

SOFTWARE
ArcGIS Image Analyst,
ArcGIS Pro Intelligence

PREDICTING LOCUST SWARM MOVEMENTS

Intergovernmental Authority on Development (IGAD)
Climate Prediction and Applications Centre (ICPAC)
Ngong, Kenya
By Kenneth Mwangi

CONTACT
Kenneth Mwangi
kenneth.mwangi@igad.int

SOFTWARE
ArcGIS Desktop, ArcGIS Spatial Analyst, Microsoft
PowerPoint, Canva

DATA SOURCES
ICPAC, Weather Research and Forecasting Model (WRF),
Copernicus Land Services, FAO Desert Locust Watch

This risk map was based on studies that identified desirable
habitats for desert locusts, examined locust reproduction
factors, and analyzed the meteorological conditions that
influence large-scale locust movement. The map is a result
of modeling these factors and was used to advise officials
during a recent locust invasion in Kenya.

Wind direction and speed often determine desert locust
movements—these factors played an important role in the
spatial analysis that predicted swarm patterns. Although
intermittent multidirectional winds and local topography
can make locusts move in varying directions, the swarms
often moved with the prevailing winds during the 2018–2021
locust outbreak.

Courtesy of IGAD Climate Prediction and Applications Centre.

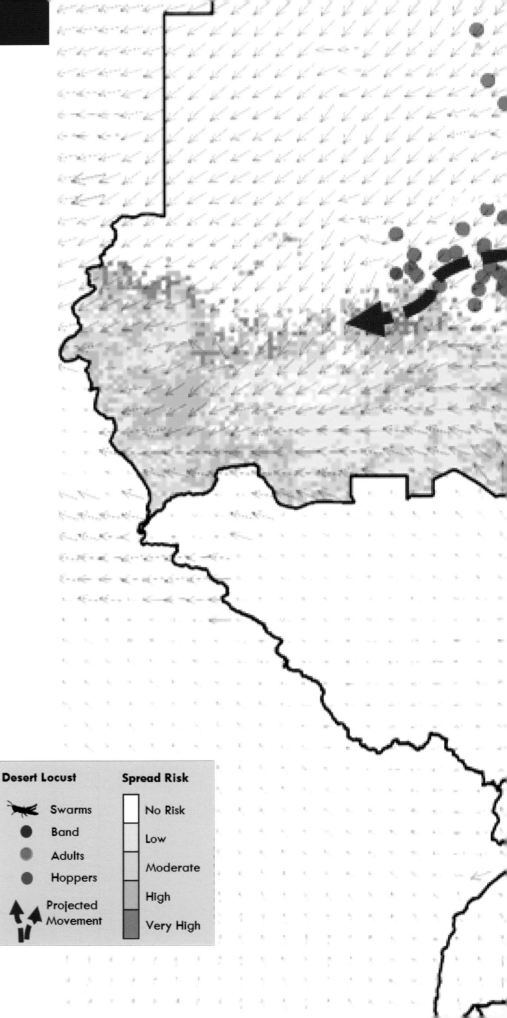

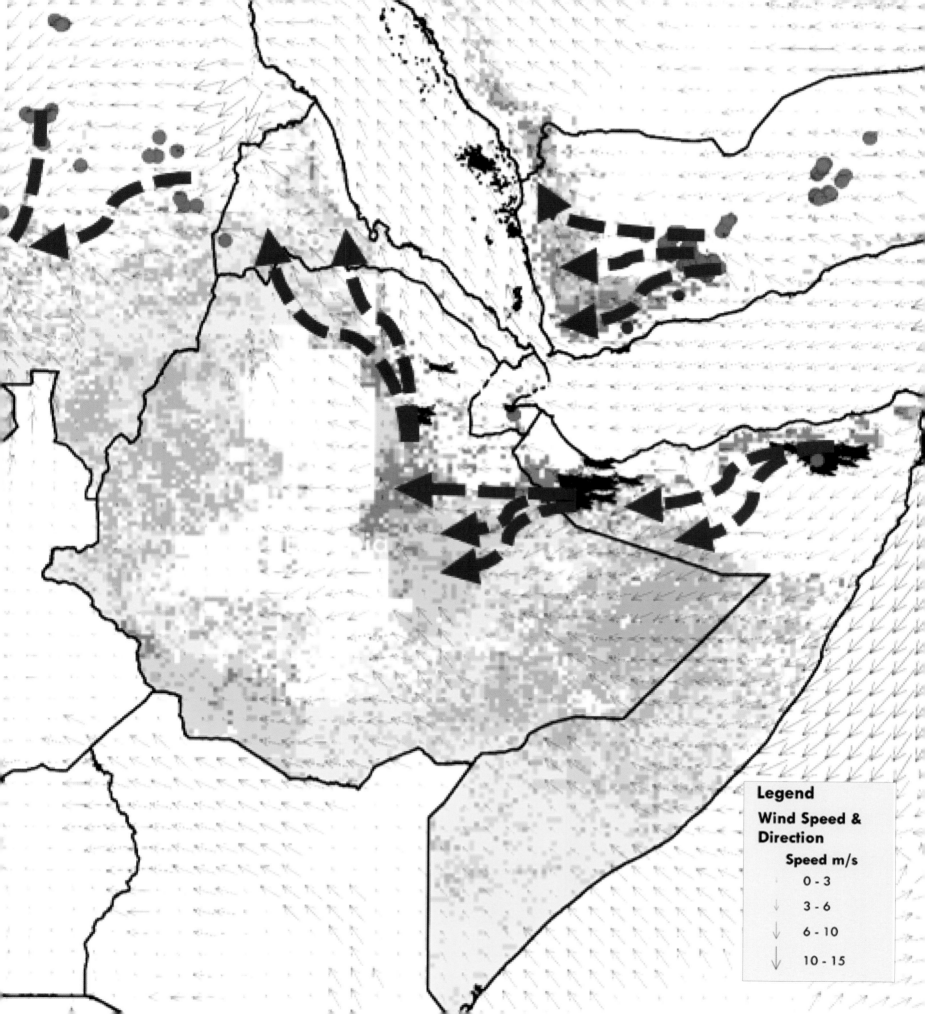

Legend

Wind Speed & Direction

Speed m/s

0 - 3

3 - 6

6 - 10

10 - 15

SOLUTIONS FOR FOREST HARVESTING

Lim Geomatics Inc.
Alberta, Canada
By Kevin Lim

CONTACT
Kevin Lim
kevin@limgeomatics.com

SOFTWARE
ArcGIS Pro, ArcGIS API for JavaScript, ArcGIS Enterprise, ArcGIS Runtime SDK for .NET

DATA SOURCES
Millar Western Forest Products Ltd., Lim Geomatics Inc.

This mobile and web solution allows easy tracking and management of forestry harvesting machines. Data can be fed into forest management tools and operational planning systems for monitoring and resource scheduling. Boundary alarms can be set to keep operators harvesting in the right areas, lidar data can be reviewed in the field, and geolocated notes can be made on-site and relayed to support staff. The web app provides data analytic tools to calculate depletion and productive machine hours and provide automated reports of other activities.

Courtesy of Lim Geomatics Inc.

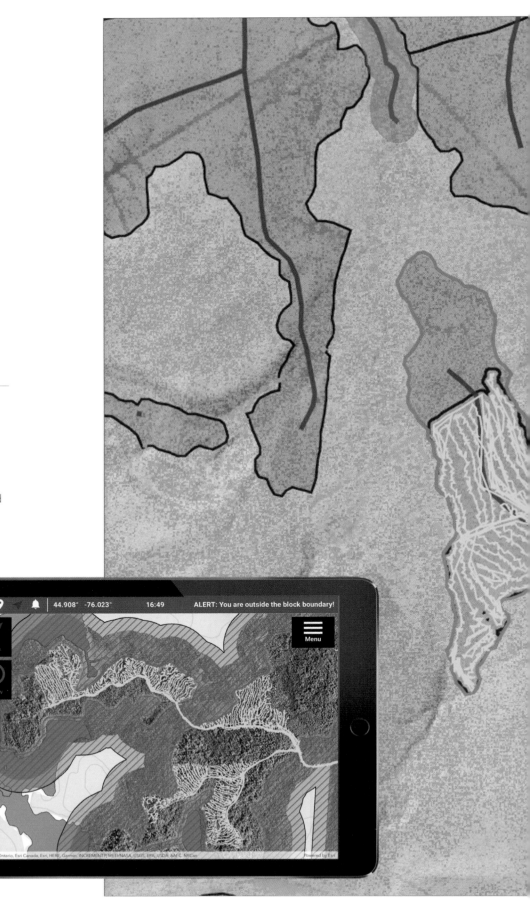

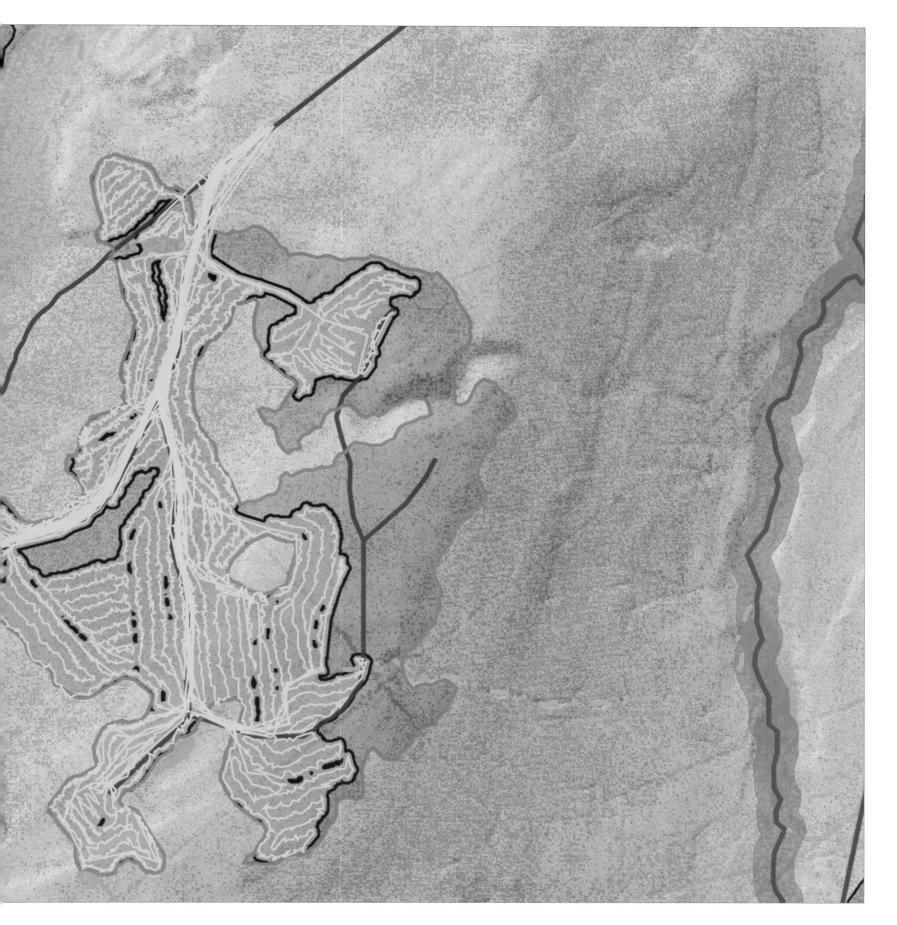

VISUALIZING THE PATTERNS OF 911 CALLS

City of Columbus
Columbus, Ohio, USA
By Robert Ford

CONTACT
Robert Ford
raford@columbus.gov

SOFTWARE
ArcGIS Pro

DATA SOURCES
Columbus Division of Fire 911: June 4, 2017, to June 4, 2022

This space-time cube was created using five years of 911 calls and shows where ladder trucks were needed in Columbus, Ohio. The analysis highlights areas in which hot spots of 911 calls are emerging and intensifying. Sporadic hot spots are also shown—areas that have been hot spots on and off over time but may not require an immediate redirection of resources. The results of this map can be used to determine service area and station distribution needs.

Courtesy of City of Columbus.

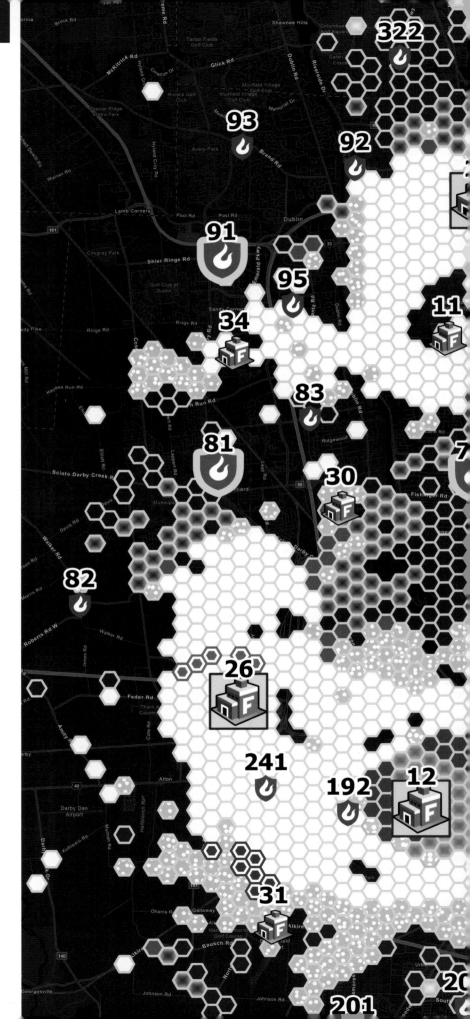

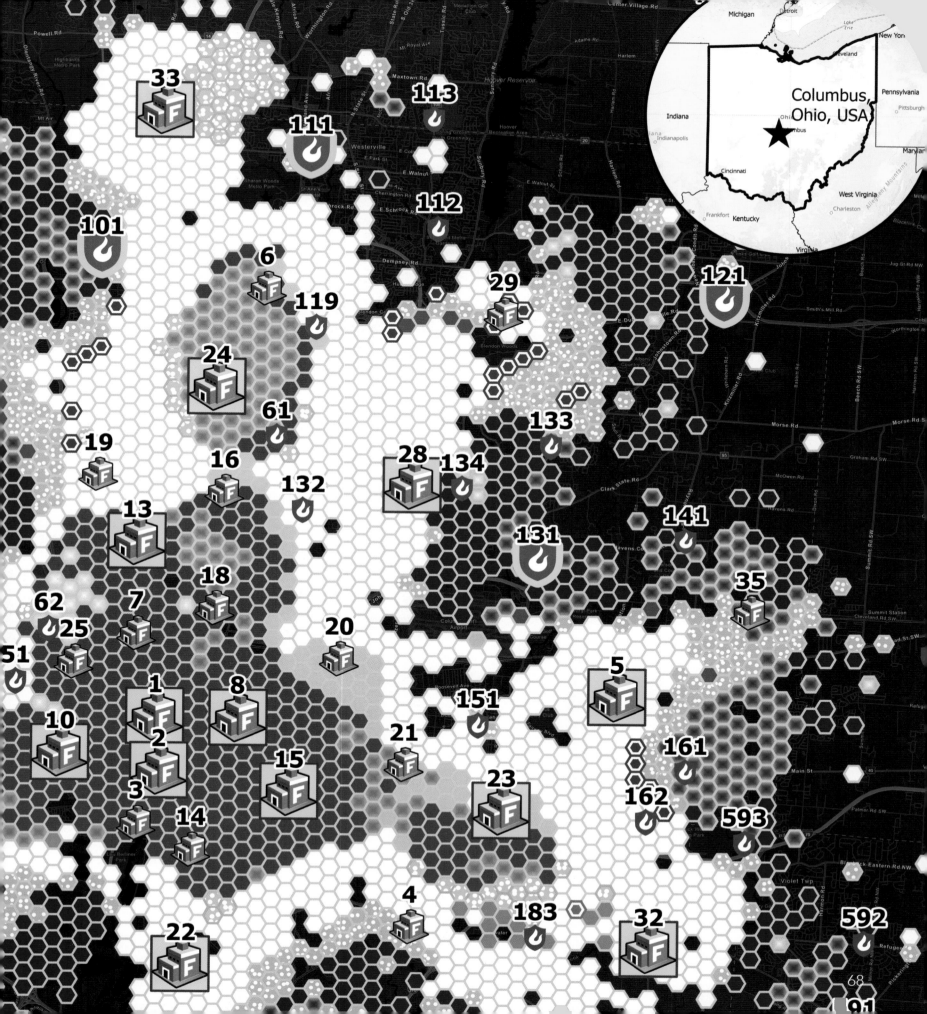

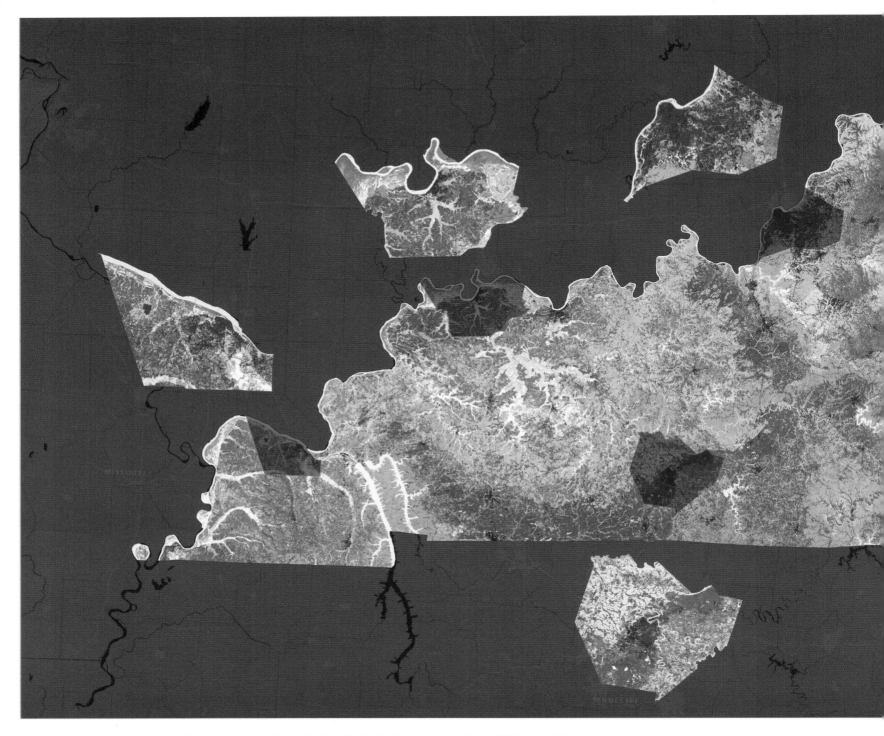

SUITABILITY MODELING FOR GREEN INFRASTRUCTURE

US Army Corps of
Engineers
Kentucky, USA
By Rachel Byrd

Green infrastructure incorporates natural processes into engineered systems to provide flood, fire, and drought risk reduction by adding capacity, flexibility, and resilience. In addition to environmental benefits, green infrastructure provides significant social benefits, including increased green space in historically underserved communities.

The goal of this project was to identify potential areas in Kentucky for implementing green infrastructure and open space designs. Five focus regions were selected that captured the diversity of physiographic landscapes and population density. A model was then developed with variables that included land cover, ecoregion, geology, wetland presence, land use, topography,

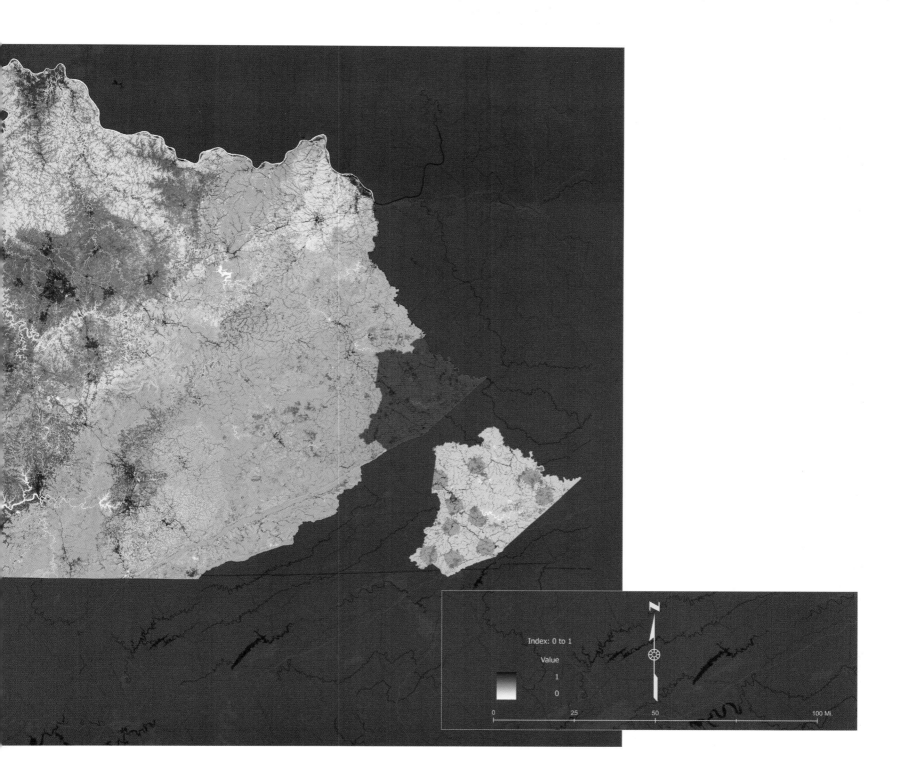

floodplains, groundwater depth, surface permeability, and karst surface presence. Next, a suitability analysis was conducted, weighting and ranking the inputs. The study results are intended to be incorporated into the Kentucky State Hazard Mitigation Plan.

Courtesy of US Army Corps of Engineers.

CONTACT
Rachel Byrd
Rachel.n.byrd@usace.army.mil

SOFTWARE
ArcGIS Pro

DATA SOURCES
Esri, Federal Emergency Management Agency (FEMA), US Department of Homeland Security, USGS, Multi-Resolution Land Characteristics Consortium (MRLC), Natural Resources Conservation Service (NRCS), US Department of Agriculture (USDA), US Forest Service (USFS), Kentucky Silver Jackets, US Army Corps of Engineers

IMPACTS FROM THE CUMBRE VIEJA VOLCANO

Takatoa
Bavaria, Germany
By Roland Degelmann

CONTACT
Roland Degelmann
rode@takatoa.net

SOFTWARE
ArcGIS Pro, ArcGIS Online, ArcGIS Notebooks

DATA SOURCES
Copernicus Emergency Management Service,
EMSR546/EMSN112

Preceded by a series of earthquakes that started on
September 11, 2021, the Cumbre Vieja volcano fully
erupted at 3:15 p.m. on September 19 the same year. The
eruption, located on the volcanic ridge of the southern half
of La Palma in the Canary Islands, lasted until December
13, 2021. The lava illustrated in the map covered more than
1,000 hectares (more than 2,400 acres) and was more than
three kilometers wide and six kilometers long, eventually
reaching the sea. It destroyed more than 3,000 buildings
and forced the evacuation of about 7,000 people.

The highlighted regions show the outlines of the lava flows
and are taken from the Emergency Management Service
(EMS) of the Copernicus Earth observation program. The
map provides a daily prediction of the impacts of future
lava flows.

Courtesy of Takatoa.

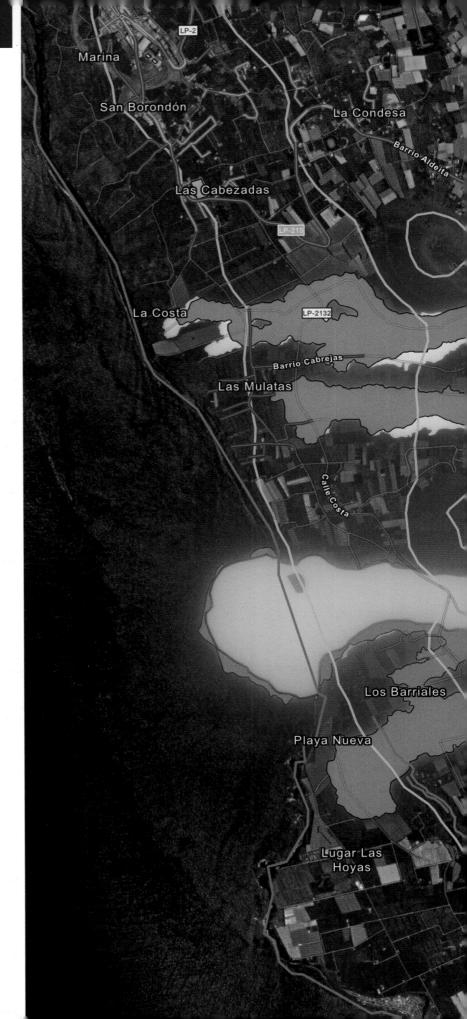

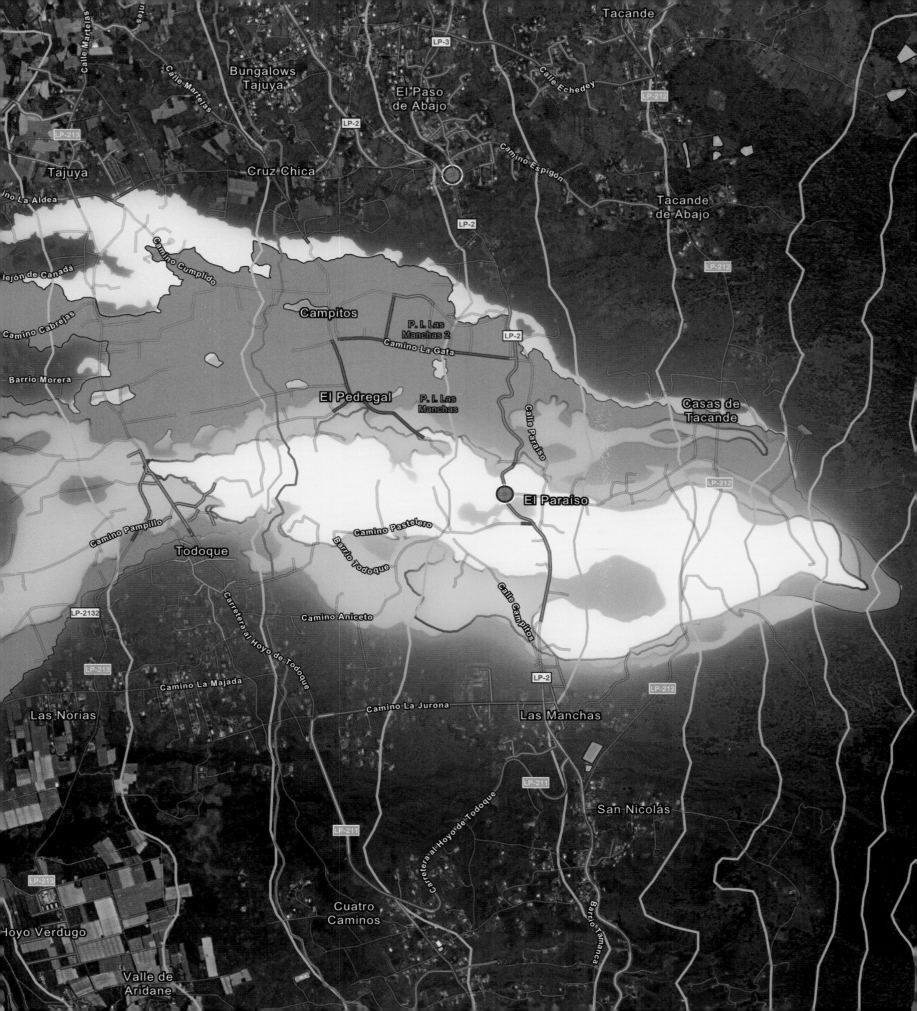

MONITORING AND RESPONDING TO FLOODS

eThekwini Metropolitan Municipality
Durban, South Africa
By Zweli Gwala

In April 2022, after weeks of torrential rains, Durban, South Africa, and its surroundings were hit by the worst floods in living memory. This dashboard provided real-time status reporting to aid data-driven disaster response. Various flood-related spatial and nonspatial datasets were integrated, detailing infrastructure impacts, such as damage to roads, drinking water, and electricity. The dashboard also included details of flood victims—where they were sheltered, potential injuries, and fatalities. Using the dashboard, city leadership and communities could better understand what was happening on the ground and respond quickly.

Courtesy of eThekwini Metropolitan Municipality.

CONTACT
Zweli Gwala
zwelibanzi.gwala@durban.gov.za

SOFTWARE
ArcMap, ArcGIS Dashboards,
ArcGIS Data Interoperability

DATA SOURCES
eThekwini Municipality, Maxar Technologies,
Council of Scientific and Industrial Research (CSIR)

FIRE STATION RESPONSE PATTERNS

City of Sugar Land
Sugar Land, Texas, USA
By Trevor Surface

CONTACT
Trevor Surface
tsurface@sugarlandtx.gov

SOFTWARE
ArcGIS Pro, ArcGIS Spatial Analyst, ESO

DATA SOURCES
ESO fire and EMS call data

In 2021, the Sugar Land Fire Department in Texas responded to more than 17,000 calls, ranging from medical matters to structure fires. Ideally, fire stations would respond to calls in their designated district, but in times of high call volume and limited availability, stations often must respond to incidents outside their districts.

For the department's annual report, the City of Sugar Land typically published call data as a heat map, but this only showed one part of the story—call location and density. For the 2021 report, a new map was developed that showed which stations responded to which calls. By connecting the responding fire station to the call location, the map illustrated the link between the volume of calls per station and the distance traveled. These origin-destination lines overlaid each fire district, allowing the fire chief to explain the full pattern of call responses and gain support for resource needs, such as increasing staffing and adding fire stations.

Courtesy of City of Sugar Land.

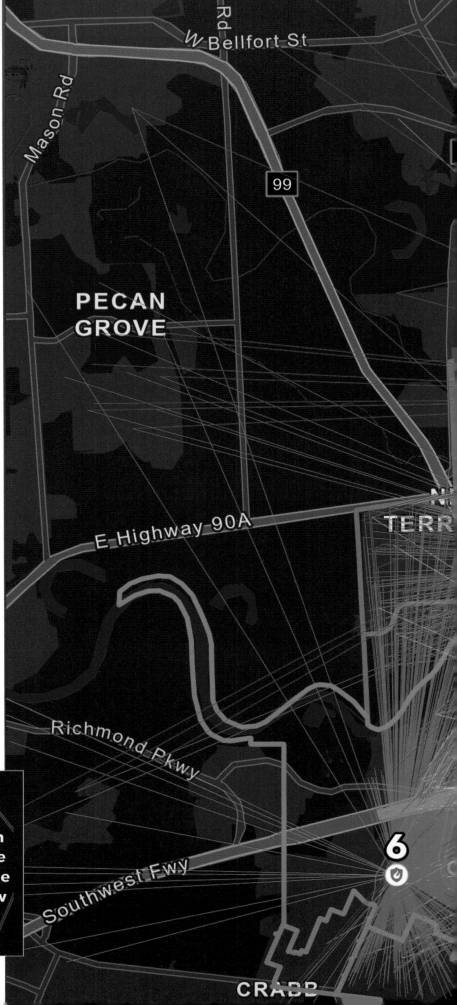

Color Guide:
Station 1 - Blue
Station 2 - Green
Station 3 - Purple
Station 4 - Orange
Station 5 - Yellow
Station 6 - Pink
Station 7 - Red

WILDFIRE MAPS FOR WESTERN COLORADO

Delta County, Colorado
Hotchkiss, Colorado, USA
By Carrie Derco

CONTACT

Carrie Derco
cderco@deltacounty.gov

SOFTWARE

ArcGIS Pro

DATA SOURCES

Colorado State Forest Service Wildfire Reduction Planner,
EagleView lidar, Quantum Spatial, Delta County

Wildfire is a major concern for property owners in western
Colorado. The ongoing megadrought and hotter-than-
average daily temperatures have drastically increased
wildfire potential. Property owners must be aware of the
risks and hazards on their land and surrounding areas.

This series of maps uses data from the Colorado Forest
Service Wildfire Reduction Planner. The map series was
shared with the public at a local fire district open house.
Coloradans learned about wildfire potential on or near
their property and spoke to firefighters about how to
mitigate wildfire impacts and plan exit routes—while
referring to the maps.

Each map in the series focuses on one of four areas of
interest: wildfire risk (shown here), fire intensity, flame
length, and rate of spread. All maps include lidar-derived
hillshading, local roads, address points, and imagery
captured at nine-inch resolution.

Courtesy of Delta County, Colorado.

FINDING REGIONAL GROWTH PATTERNS FROM NEW CONSTRUCTION PERMITS

City of Houston
Planning & Development Department
Houston, Texas, USA
By Sona Sunny

CONTACT
Sona Sunny
sona.sunny@houstontx.gov

SOFTWARE
ArcGIS Pro

DATA SOURCES
City of Houston GIS Database City of Houston Permitting Center, Houston Public Works

Houston is the fourth-largest city in the US, with more than two million residents, spread out over 671 square miles. Tracking the number of permits issued for new construction can be used to measure regional growth trends. This map illustrates the spatial and temporal growth trends from 2015 to 2020. The city's permitting office stores permit data in tables, which are analyzed to find out where growth is happening and provide important information about trends.

Courtesy of City of Houston.

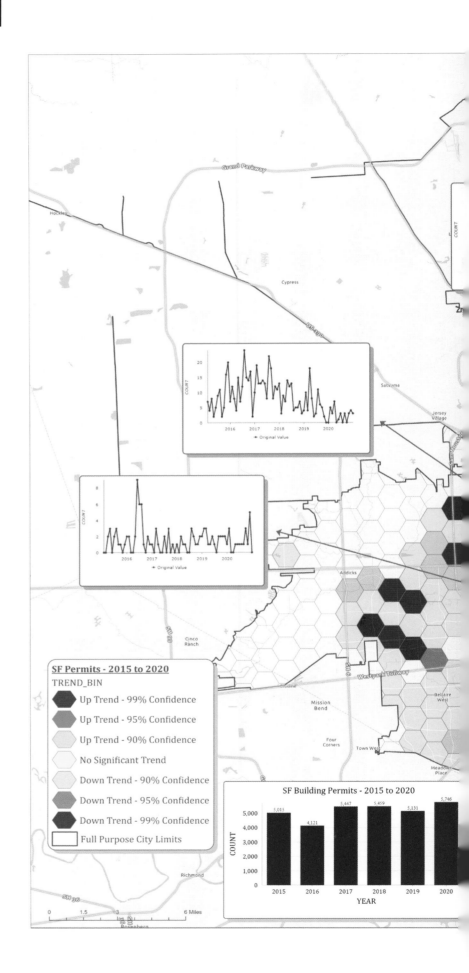

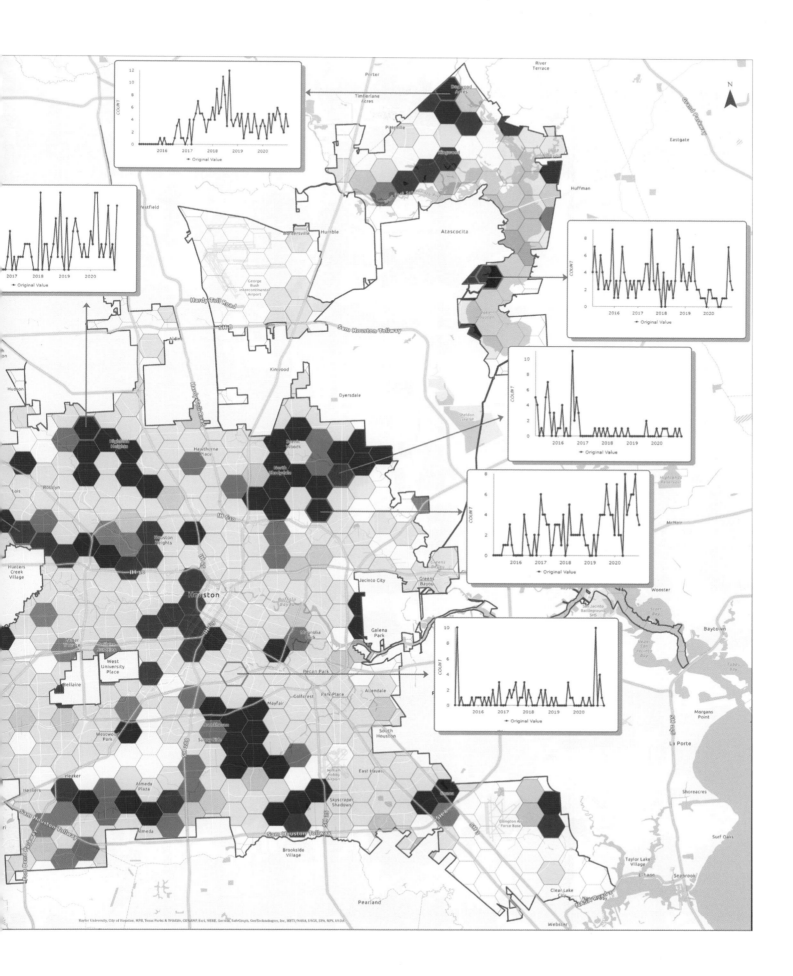

VILLAGE DEVELOPMENT PATTERNS OVER TIME

County of Waukesha
Wales, Wisconsin, USA
By Jim Landwehr

CONTACT
Jim Landwehr
jlandwehr@waukeshacounty.gov

SOFTWARE
ArcGIS Pro

DATA SOURCES
Wisconsin Land Information Program

This map was created for display at the centennial celebration for the village of Wales, Wisconsin. It depicts plat development over time, beginning with the railway depot in the late 1800s, and illustrates the expanding pattern of village development. The date of each plat was included when the database was developed, so there is a rich temporal dataset that makes this and other time-driven development maps easy to create.

Courtesy of County of Waukesha.

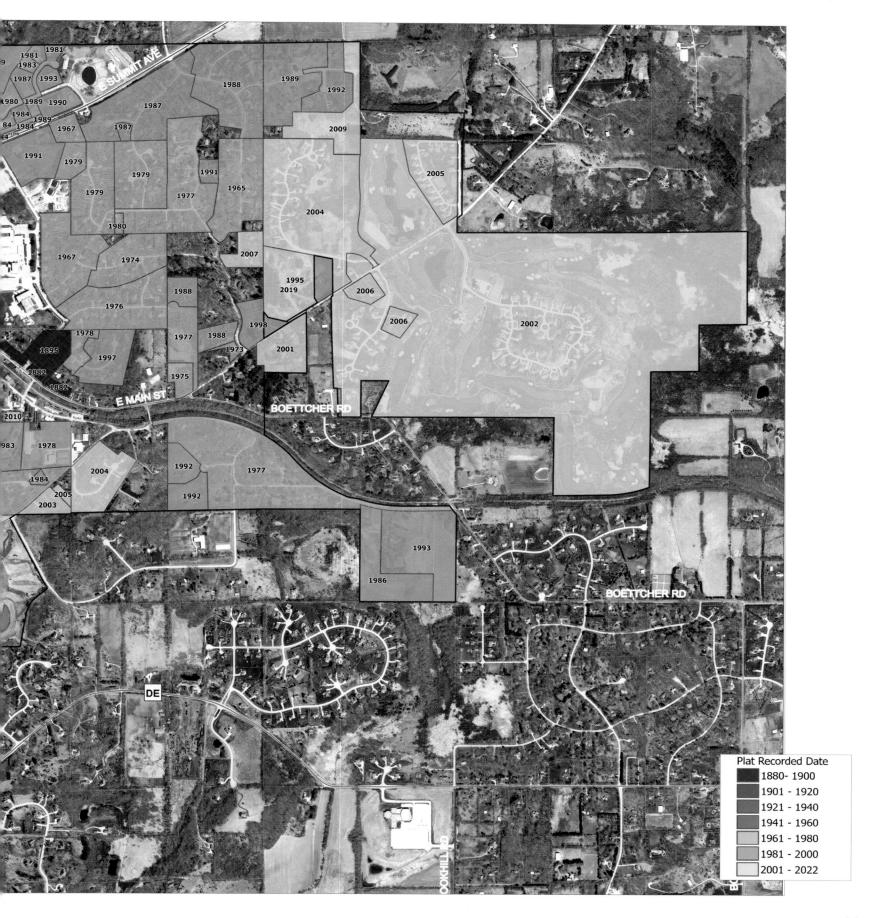

Plat Recorded Date
- 1880- 1900
- 1901 - 1920
- 1921 - 1940
- 1941 - 1960
- 1961 - 1980
- 1981 - 2000
- 2001 - 2022

ZONING MAPS IN 3D

Columbus Consolidated Government
Columbus, Georgia, USA
By David Cooper

CONTACT
David Cooper
davidcooper@columbusga.org

SOFTWARE
ArcGIS Pro

DATA SOURCES
Columbus Consolidated Government; ArcGIS Living Atlas of the World; Airbus; USGS; NGA; NASA; CGIAR; NCEAS; NLS; OS; NM; the GIS User Community, Columbus

This 3D visualization of buildings indicates the zoning restrictions along a stretch of highway in Columbus, Consolidated Government.

Courtesy of Columbus Consolidated Government.

LAND USE IN THE UNITED ARAB EMIRATES

American University of Sharjah
Sharjah, United Arab Emirates
By Rahul Gawai

CONTACT
Rahul Gawai
rgawai@aus.edu

SOFTWARE
ArcGIS Pro, ArcGIS Living Atlas of the World

DATA SOURCES
Esri, ArcGIS Living Atlas, European Space Agency
(ESA) Sentinel-2 satellite imagery

The United Arab Emirates is one of the fastest
developing nations in the world. Monitoring
changes to land use and land cover is an important
aspect of planning developments in the region.
Almost 65 percent of land in this region is classified
as bare or desert land, with an additional 30 percent
of land classified as scrub. Just 4.5 percent is
considered built area, whereas the remaining tiny
fraction of land falls under the remaining categories.
This land-use and land-cover map used a layer—
developed through a machine learning workflow—
that analyzed global land cover with 10 meters of
spatial resolution.

Courtesy of American University of Sharjah.

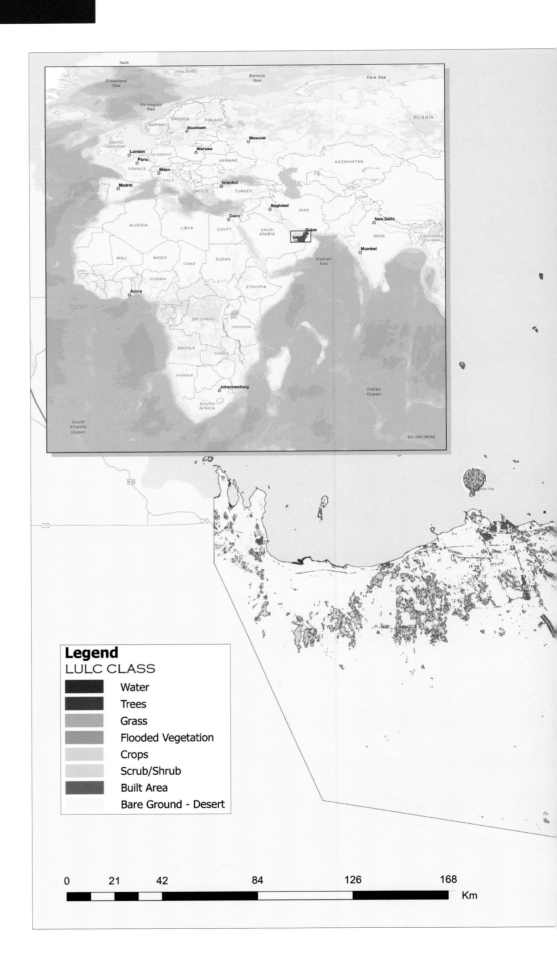

Legend
LULC CLASS
- Water
- Trees
- Grass
- Flooded Vegetation
- Crops
- Scrub/Shrub
- Built Area
- Bare Ground - Desert

0   21   42   84   126   168
Km

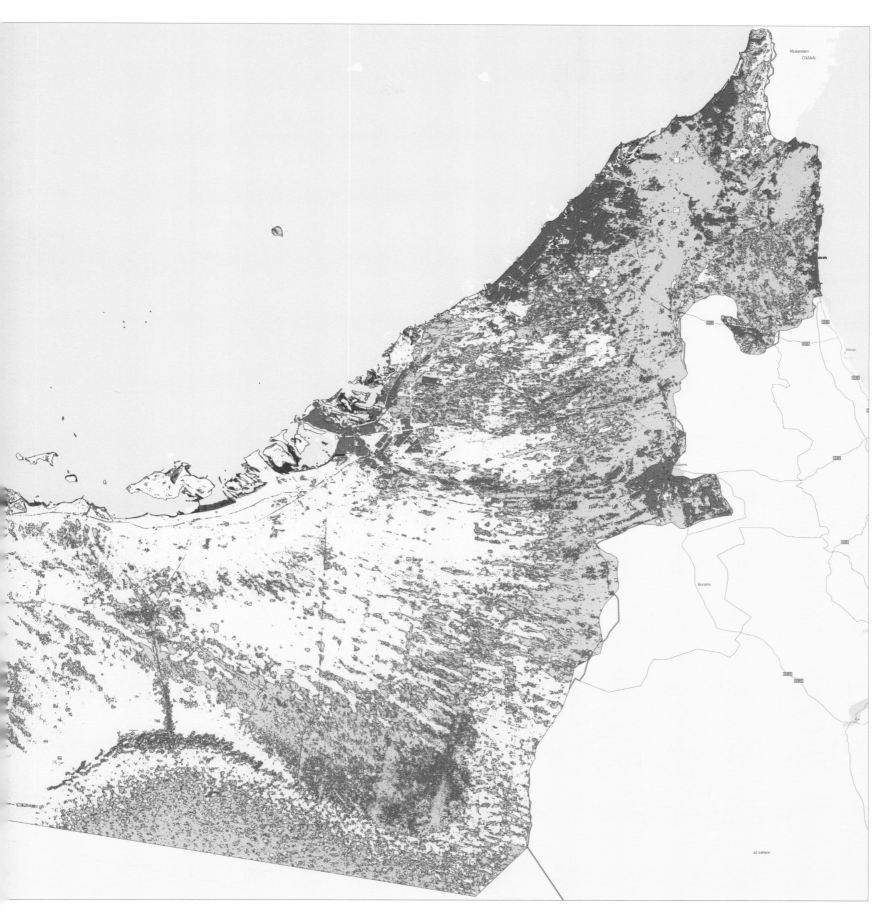

CITY OF TORRANCE ELECTORAL DISTRICTS

City of Torrance
Torrance, California, USA
By Sunny Lai

CONTACT
Sunny Lai
mlai@torranceca.gov

SOFTWARE
ArcGIS Pro

DATA SOURCES
City of Torrance, Los Angeles County Assessor

On June 19, 2018, the Torrance, California, City Council adopted an ordinance establishing by-district elections for council offices and approved an election calendar for the newly created districts. The City of Torrance was divided into six districts based on population.

The Torrance Electoral Districts map series provides city council candidates and Torrance residents with an accurate boundary of each district, zoning details, schools, parks, libraries, community and recreation centers, and critical features within the district boundary.

Courtesy of City of Torrance.

SCHOOLS

Elementary School (5)
Elementary School Prvt (2)
Middle School (1)
High School (1)
College (1)
Adult School (1)

P Police Stations (1)
Fire Stations (1)

Hawthorne Blvd Overlay District

City Boundary
Railroad

Total Sqaure Miles: 3

COMMUNITY / RECREATION CENTERS

**Center Name**

15 Herma Tillim Center
20 The Roadium Open Air Market
23 Alondra Aquatic Center
24 North Torrance Community Center

PARKS

Park Name

DESCANSO
GUENSER
LA CARRETERA
McMASTER
OSAGE

ZONING

	(10)
	R3 (4)
	A1 (17)
	A1/M1/M2 (1)
	A1/R2 (7)
	C2 (96)
	C2-PP (3)
	C2/P1 (2)
	C2/R3 (1)
	C3 (15)
	C3-PP (13)
	C5 (1)
	CR (26)
	H-NT (13)
	M1 (27)
	M2 (2)
	ML (9)
	PU (17)
	R-1 (3)
	R1 (5,620)
	R1-PP (6)
	R2 (117)
	R3 (452)
	R3-PP (205)
	R3/C2 (1)
	R3/R1 (8)
	R4 (49)
	R5 (1)
	RP (5)
	RP-PP (21)
	RR3-PP (51)

TAX ASSESSMENTS IN 3D

Houseal Lavigne Associates
Fort Wayne, Indiana, USA
By Sujan Shrestha and Brandon Nolin, Houseal Lavigne

CONTACT
Devin Lavigne
dlavigne@hlplanning.com

SOFTWARE
ArcGIS Pro

DATA SOURCES
Office of the Allen County Assessor, US Census Bureau, Houseal Lavigne

The tax revenue generated by development is a key consideration in ensuring that land use policy supports fiscally responsible growth. To document this, the assessed values of properties throughout Allen County were aggregated at the census block level and mapped to identify development with high assessed property tax value per acre. Higher assessed values are generally found in areas that are built at higher densities within municipal boundaries and served by water and sewer infrastructure. Downtown Fort Wayne is clearly the highest value district within the region, making up only .08 percent of Allen County's land area, but accounting for 2 percent of all assessed value (a 25:1 ratio despite including tax-exempt properties). The analysis also helped identify undervalued areas where neighborhood built form is similar to higher value neighborhoods, indicating untapped development potential in well-served areas. This information was integral to developing a plan for regional growth and development as part of the All in Allen comprehensive planning process.

Courtesy of Houseal Lavigne Associates.

Assessed Value Per Acre

Assessed value per acre by block. Each foot of vertical elevation represents $1,000 in assessed tax value.

Less than or equal to $75,000
$75,000 - $150,000
$150,000 - $300,000
$300,000 - $500,000
$500,000 - $1,000,000
$1,000,000 - $3,000,000
Greater than $3 million

Block Assessed Value
+ Indiana Power

Block Assessed Value - $1.3 m
+ Lutheran Hospital

Block Assessed Value - $0.86 mil
+ Foster Park

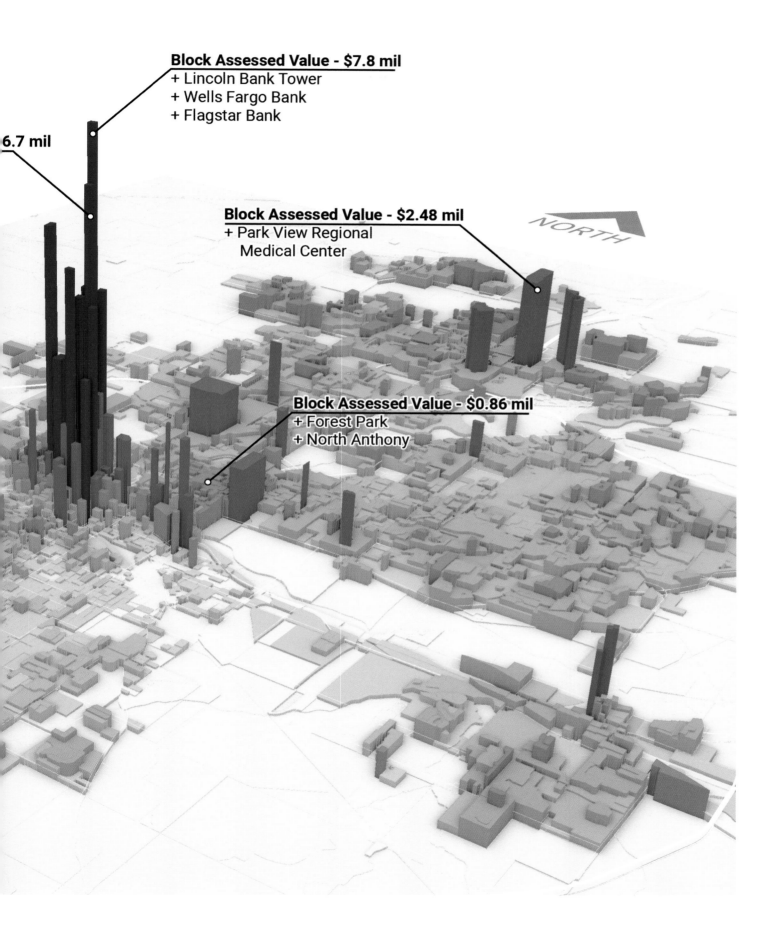

Block Assessed Value - $7.8 mil
+ Lincoln Bank Tower
+ Wells Fargo Bank
+ Flagstar Bank

6.7 mil

Block Assessed Value - $2.48 mil
+ Park View Regional
  Medical Center

NORTH

Block Assessed Value - $0.86 mil
+ Forest Park
+ North Anthony

SAN JUAN NATIONAL FOREST VISITOR MAP

USDA Forest Service
Colorado, USA
By Jesse Nett, Mark Roper, and the San Juan National Forest

CONTACT
Jesse Nett
jesse.nett@usda.gov

SOFTWARE
ArcGIS Pro, ArcMap, Adobe Creative Cloud,
Avenza MAPublisher, Geographic Imager

DATA SOURCES
USDA Forest Service, Bureau of Land Management (BLM),
State of Colorado, Esri

The 1.9 million acres of San Juan National Forest in southwestern Colorado, on the western slope of the Continental Divide, are designated by Congress as a national forest and provide for a range of recreational activities, resource uses, and habitat needs. This national forest is for everyone to enjoy.

The visitor map illustrates recreational opportunities, travel routes, and points of cultural and historical interest for visitors to explore.

Courtesy of USDA Forest Service.

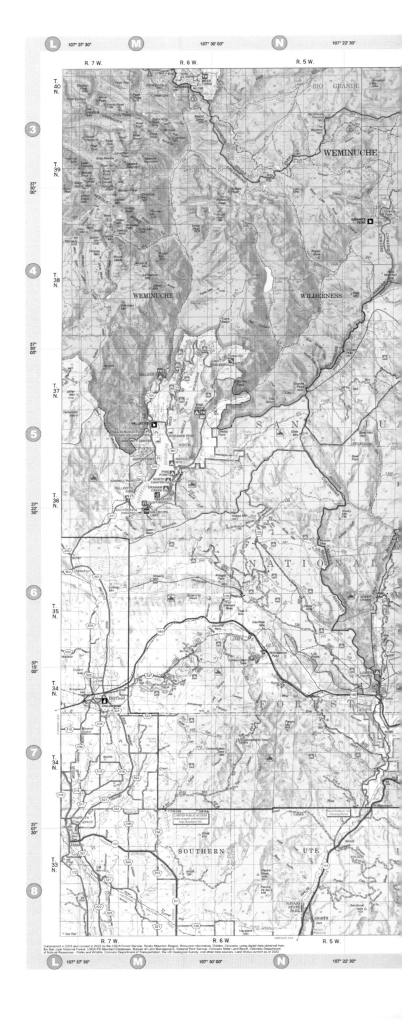

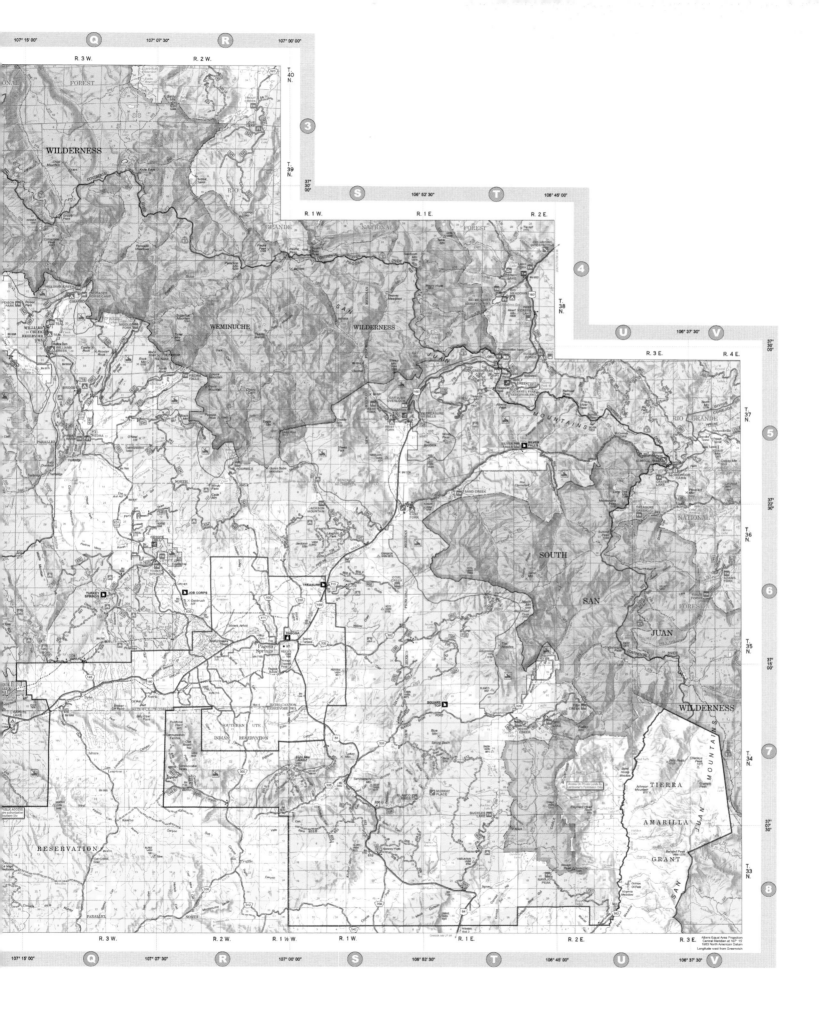

NEW YORK BOTANICAL GARDEN PLANTTRACKER

Blue Raster LLC
Arlington, Virginia, USA
By Blue Raster

CONTACT
Christopher Gabris
cgabris@blueraster.com

SOFTWARE
ArcGIS Enterprise, ArcGIS Workforce,
ArcGIS Field Maps, BG-BASE

DATA SOURCES
New York Botanical Garden

PlantTracker is the online catalog of the New York
Botanical Garden's (NYBG) diverse living collections of
more than one million plants from around the world set in
50 gardens, collections, and across its 250-acre National
Historic Landmark landscape in the Bronx, New York.
This mobile-friendly mapping application allows visitors,
students, and scientists to find out which plants NYBG
grows and to learn about them through brief descriptions
and photographs.

Courtesy of Blue Raster LLC.

Bronx River Pkwy

Bronx River Pkwy

pelham Pkwy

E Fordham Rd

Dr Theodore Kazimiroff Blvd

Quercus rubra

Common Name: northern red oak
Accession ID: 939/96*B
Family: FAGACEAE
Habit: tree
Native: Nova Scotia to Georgia, w. to Minnesota & Oklahoma
Hardiness: USDA Zone 3
Light: sun, partial sun
Location: Rock Garden - Azalea Woods

Get Details

MAINTAINING THE DEEPWATER PORTS OF TEXAS

US Army Corps of Engineers
Galveston, Texas, USA
By Brittany Tiemann

CONTACT
Brittany Tiemann
brittany.l.tiemann@usace.army.mil

SOFTWARE
ArcGIS Pro, ArcGIS Online

DATA SOURCES
The Britannica Dictionary definitions, the Organisation for Economic Co-operation and Development (OECD) Glossary of Statistical Terms, Esri Oceans vessel traffic data

These deepwater ports of Texas are shown with their authorized channel depths, as maintained by the US Army Corps of Engineers in partnership with regional entities and industry; their future depths; and the current trend of vessels visiting each port. This map provides insight into the future of Texas port traffic when channel-deepening projects are concluded and channels can facilitate deeper drafting vessels.

Courtesy of the US Army Corps of Engineers.

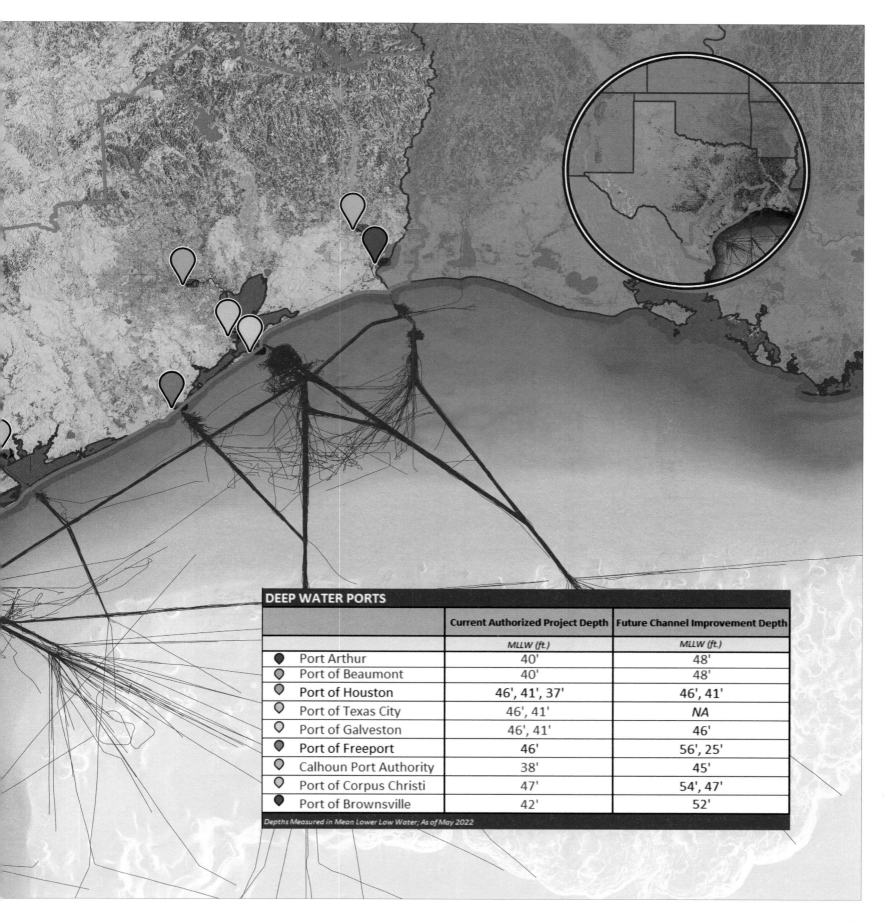

DEEP WATER PORTS		Current Authorized Project Depth	Future Channel Improvement Depth
		MLLW (ft.)	MLLW (ft.)
●	Port Arthur	40'	48'
●	Port of Beaumont	40'	48'
●	Port of Houston	46', 41', 37'	46', 41'
●	Port of Texas City	46', 41'	NA
●	Port of Galveston	46', 41'	46'
●	Port of Freeport	46'	56', 25'
●	Calhoun Port Authority	38'	45'
●	Port of Corpus Christi	47'	54', 47'
●	Port of Brownsville	42'	52'

Depths Measured in Mean Lower Low Water; As of May 2022

CHARTING THE NORTHWEST PASSAGE

Tufts University
Somerville, Massachusetts, USA
By Adina Zucker

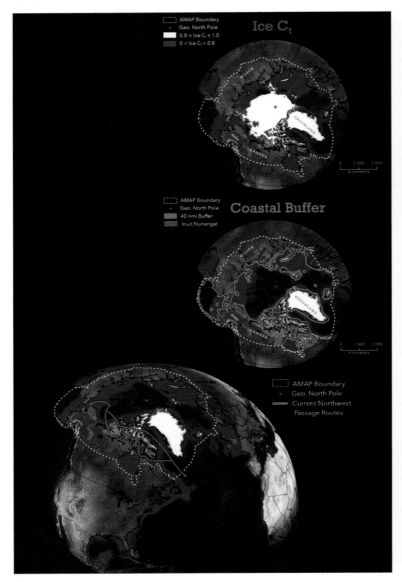

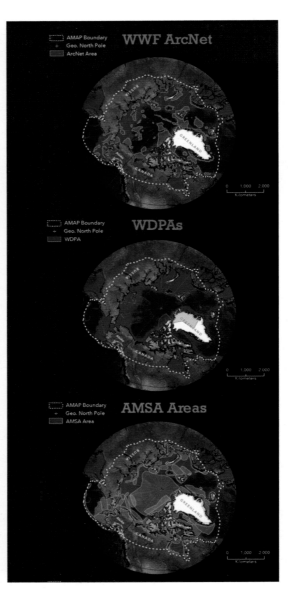

As temperatures rise in the Arctic, sea ice melts and new routes become available. Vessel traffic in this previously remote region is already causing disruptions to local wildlife, and yet shipping is projected to expand in the coming years. With traffic numbers still relatively low compared with other regions of the world, policy makers have the flexibility to establish shipping policies that protect Arctic marine wildlife.

Today vessels take six common routes, collectively known as the Northwest Passage, when traversing the Canadian Arctic. The map charts a single shipping fairway through the region for container, bulk carrier, and tanker ships. This single fairway minimizes contact with areas of heightened conservation concerns and balances economics and safety.

Courtesy of Tufts University.

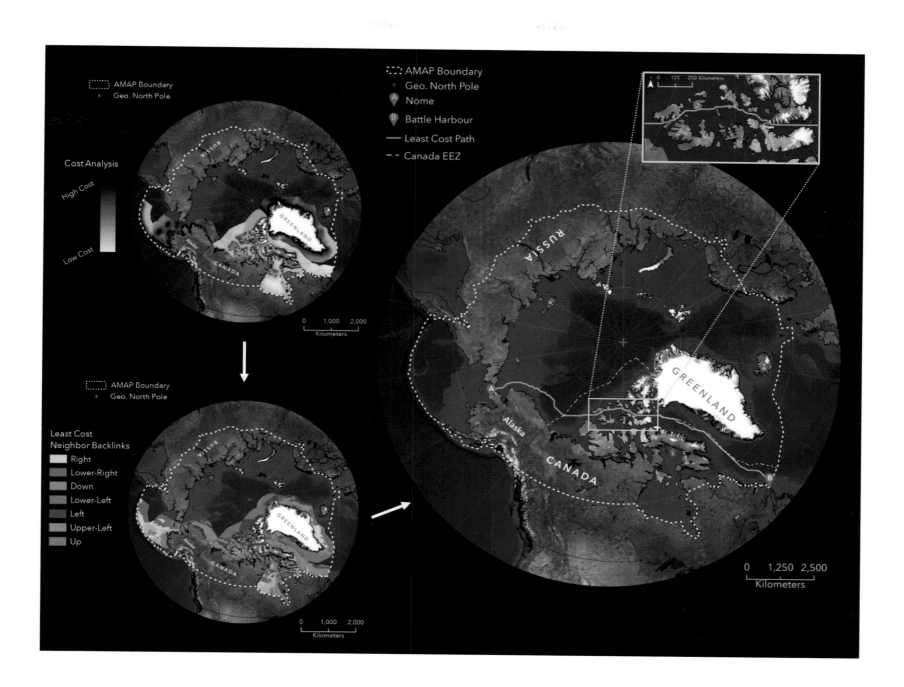

Cost Analysis

High Cost

Low Cost

AMAP Boundary
Geo. North Pole

AMAP Boundary
Geo. North Pole
Nome
Battle Harbour
Least Cost Path
Canada EEZ

Least Cost Neighbor Backlinks

Right
Lower-Right
Down
Lower-Left
Left
Upper-Left
Up

0   125   250 Kilometers

RUSSIA

GREENLAND

A

Alaska

CANADA

B

0   1,250   2,500
Kilometers

0   1,000   2,000
Kilometers

0   1,000   2,000
Kilometers

CONTACT
Adina Zucker
Adina.Zucker@tufts.edu

SOFTWARE
ArcGIS Pro

DATA SOURCES
Esri, NGA, WWF Arctic Programme, Conservation of
Arctic Flora and Fauna (CAFF), USNIC, Statistics Canada,
Government of Canada, Natural Earth, the General
Bathymetric Chart of the Oceans (GEBCO), Protection of
the Arctic Marine Environment (PAME), Protected Planet

HIGHWAY VIEWSHED ANALYSIS

County of Pickens
Pickens County, South Carolina, USA
By Jimmy Threatt and Brian Ritter

CONTACT
Jimmy Threatt or Brian Ritter
jimmyt@co.pickens.sc.us, brianr@co.pickens.sc.us

SOFTWARE
ArcGIS Pro, ArcMap, ArcGIS 3D Analyst

DATA SOURCES
North Carolina (NC) OneMap, South Carolina Forestry
Commission, Pickens County, South Carolina State GIS
Department, Greenville County, South Carolina Department
of Natural Resources

In spring 2022, the Pickens County Council adopted an
ordinance to establish standards and limitations on future
development adjacent to Highway 11. The new standards
were designed to protect the natural beauty along the scenic
byway. To aid the council, the Pickens County GIS Mapping
Department developed this viewshed analysis, identifying
unprotected areas along the road corridor.

Courtesy of County of Pickens.

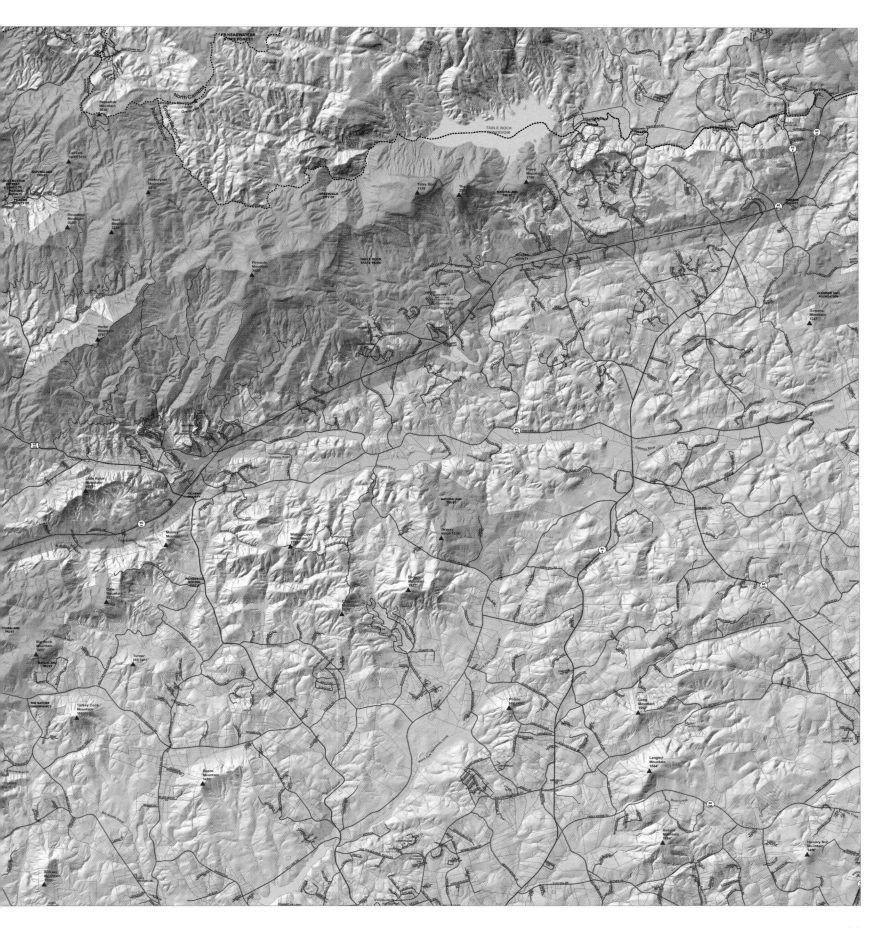

EVALUATING BUS TRANSPORTATION ROUTES

Dallas Area Rapid Transit
Dallas, Texas, USA
By Ali Behseresht

CONTACT
Ali Behseresht
abehseresht@dart.org

SOFTWARE
ArcGIS Pro, ArcGIS Network Analyst™

DATA SOURCES
Longitudinal Employer-Household Dynamics

Dallas Area Rapid Transit (DART) implemented its redesigned Bus Network in January 2022. The network was redesigned to provide DART customers a more reliable and accessible public transit network.

Network analysis was used to evaluate service to job locations by way of the new network and showed that, although the number of routes and stops was reduced, service to job locations improved.

Courtesy of Dallas Area Rapid Transit.

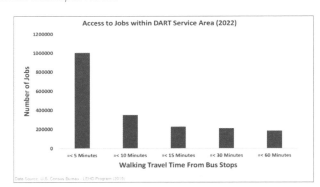

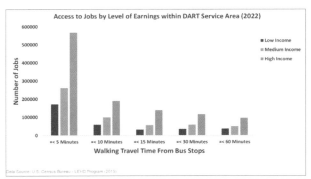

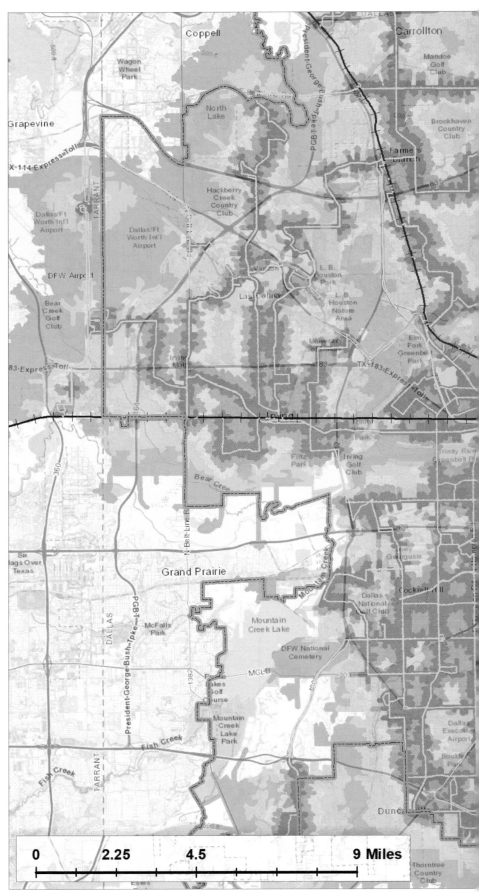

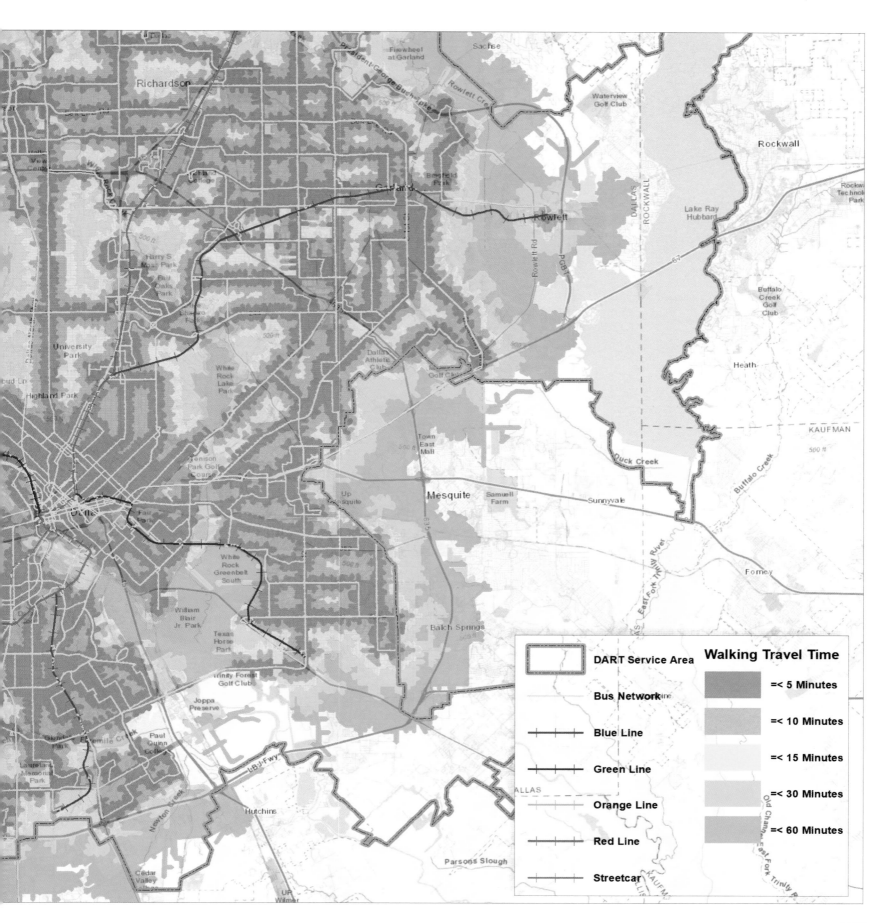

	DART Service Area	**Walking Travel Time**
	Bus Network	=< 5 Minutes
	Blue Line	=< 10 Minutes
	Green Line	=< 15 Minutes
	Orange Line	=< 30 Minutes
	Red Line	=< 60 Minutes
	Streetcar	

SPATIAL ANALYSIS OF TRAFFIC INCIDENTS IN MEXICO CITY

Avanti Engineering Group LLC
Southlake, Texas, USA
By Victoria Urueta and Sergio Lugo

CONTACT
Sergio Lugo
sergiolugo@avantieg.com

SOFTWARE
ArcGIS Pro, Microsoft Office

DATA SOURCES
Center for Command, Control, Computers, Communications, and Citizen Contact of Mexico City (C5)

Mexico City is one of the most congested metropolitan areas in the world. In 2019, an average of 674 daily traffic incidents in Mexico City were reported. C5 of Mexico City is a local agency that maintains records for all traffic incidents. These incidents are classified into various categories: with or without injuries, cyclist involved, pedestrian hit, and so on.

This spatial analysis was designed to identify geographic and temporal trends in traffic incidents and provide information to authorities, who can evaluate hot spots and implement measures to reduce the number of accidents.

Victoria Urueta and Sergio Lugo.

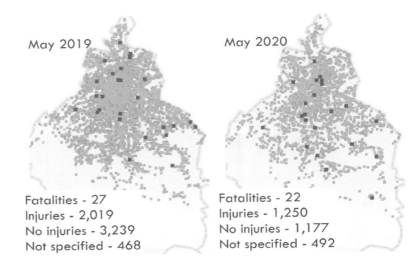

May 2019
Fatalities - 27
Injuries - 2,019
No injuries - 3,239
Not specified - 468

May 2020
Fatalities - 22
Injuries - 1,250
No injuries - 1,177
Not specified - 492

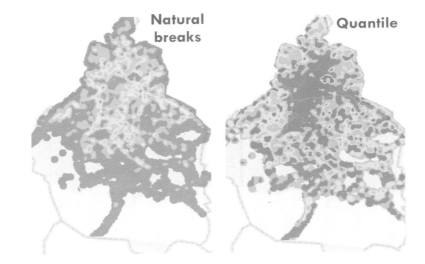

Natural breaks

Quantile

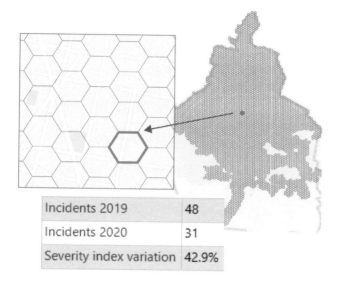

Incidents 2019	48
Incidents 2020	31
Severity index variation	42.9%

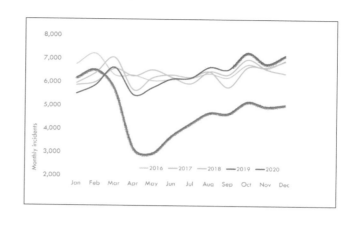

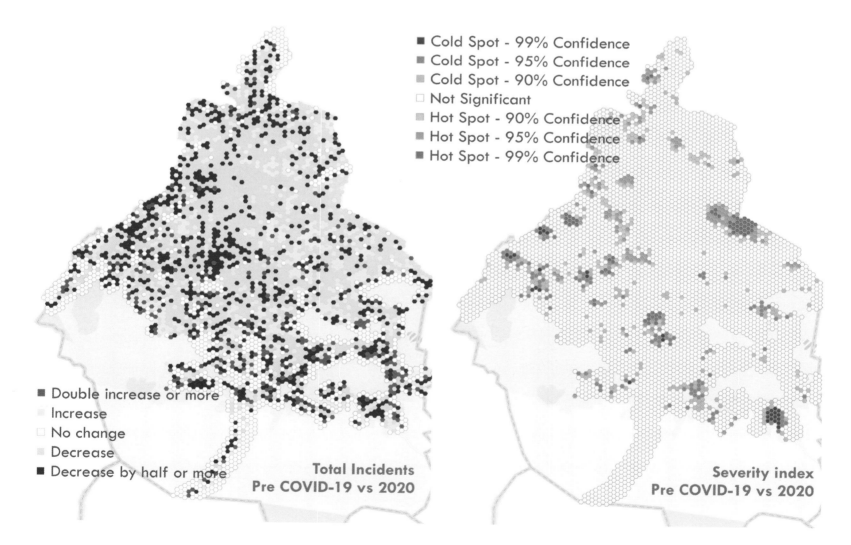

■ Cold Spot - 99% Confidence
■ Cold Spot - 95% Confidence
■ Cold Spot - 90% Confidence
□ Not Significant
■ Hot Spot - 90% Confidence
■ Hot Spot - 95% Confidence
■ Hot Spot - 99% Confidence

■ Double increase or more
■ Increase
□ No change
■ Decrease
■ Decrease by half or more

Total Incidents
Pre COVID-19 vs 2020

Severity index
Pre COVID-19 vs 2020

TRAFFIC VOLUMES BY STREET SEGMENT

Unified Government of Wyandotte County
and Kansas City, Kansas
Kansas City, Kansas, USA
By Robert Anderson

CONTACT
Robert Anderson
conrad.maps@gmail.com

SOFTWARE
ArcGIS Pro, ArcGIS Online, ArcGIS StoryMaps

DATA SOURCES
Unified Government of Wyandotte County and Kansas City,
Kansas (UG); UG Public Works; UG GeoSpatial Services;
TranSystems; StreetLight Data

This map depicts traffic volumes in 3D and is part of a
broader dataset used to inform a GIS-driven capital
improvement plan. Visualizing traffic informs staff, policy
makers, and the public of the value of each transportation
asset and helps estimate the impact of a closure. This map is
used to prioritize intersection improvements, bridge repairs,
road preservation programs, and long-range capital projects.

In 2022, two crossing bridges were shut down for failing
safety inspections. The cost to replace each bridge was
estimated at $60 million to $140 million and would consume
around five to seven times the annual capital improvement
budget. The map helped decision-makers decide which
of the two bridge-repair projects would need additional
government funding.

In the future, this map will help establish vehicle miles
traveled (VMT) policies that support a transition to a post-
fossil-fuel transportation system. Government agencies will
need this type of map to understand how VMT policies will
distribute funds and affect local maintenance funding, which
currently relies on a gas tax.

*Courtesy of Unified Government of Wyandotte County and Kansas
City, Kansas. Data designed and prepared by UG Public Works
(Brandon Grover, Jake Kooyer, Robert Anderson), TranSystems (Mike
Wahlstedt), and StreetLight Data (Jim Hubbell, Sean Co, Jon Wergin,
and Louis Yan).*

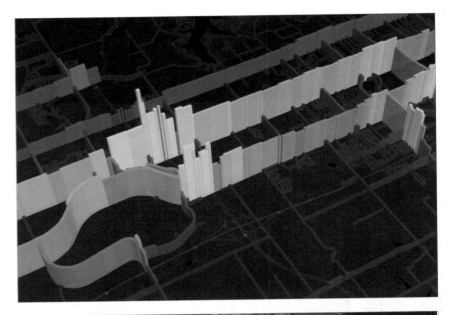

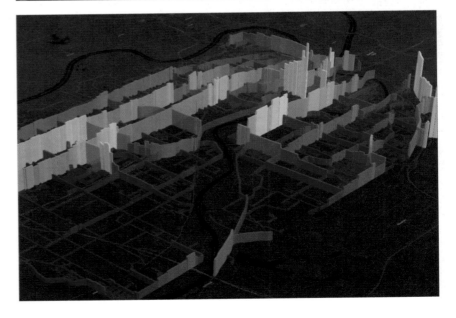

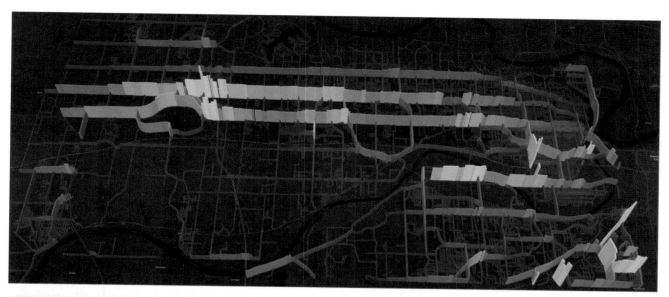

SOUTHERN METROPOLITAN REGION PRODUCTIVITY BY 2050

Department of Environment, Land, Water, and Planning (DELWP)
Melbourne, Australia
By Sara Faizan, Paula McNamara, Renee Ward, Lauren McCowan, and Tim Gray

CONTACT
Justin Madex
justin.madex@delwp.vic.gov.au

SOFTWARE
ArcGIS Pro

DATA SOURCES
DELWP

This map shows how productivity will be enhanced across the Southern Metropolitan Region of Melbourne, Australia, by 2050. The Southern Metropolitan Region is one of six regions in metropolitan Melbourne for which land-use framework plans have been completed.

The map shows the key locations of future employment or population growth, including metropolitan and major activity centers, national employment and innovation clusters, state—or regionally significant— industrial and commercial areas, and future business corridors. It shows how these locations will be connected by existing and future rail and road networks and indicates the location of transport gateways and intermodal transport hubs. It also shows the valuable agricultural land surrounding the metropolitan area and land that has been set aside for future extractive industries.

Courtesy of DELWP.

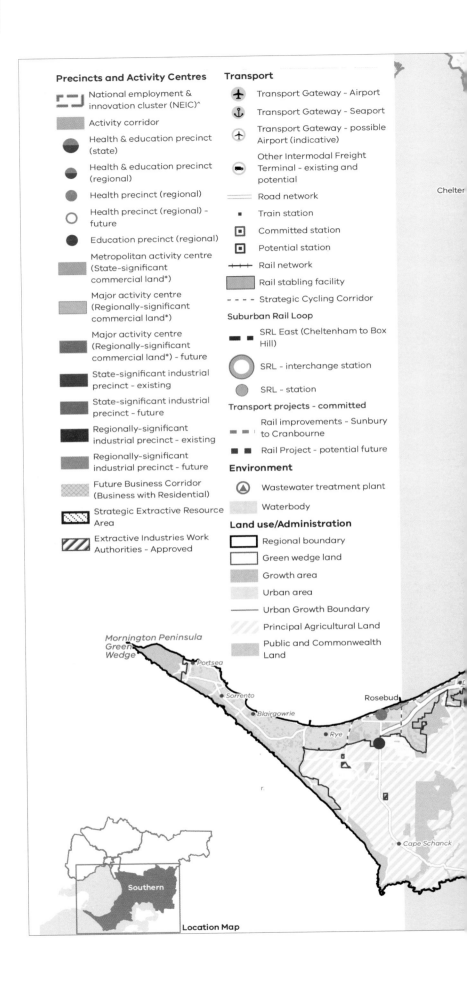

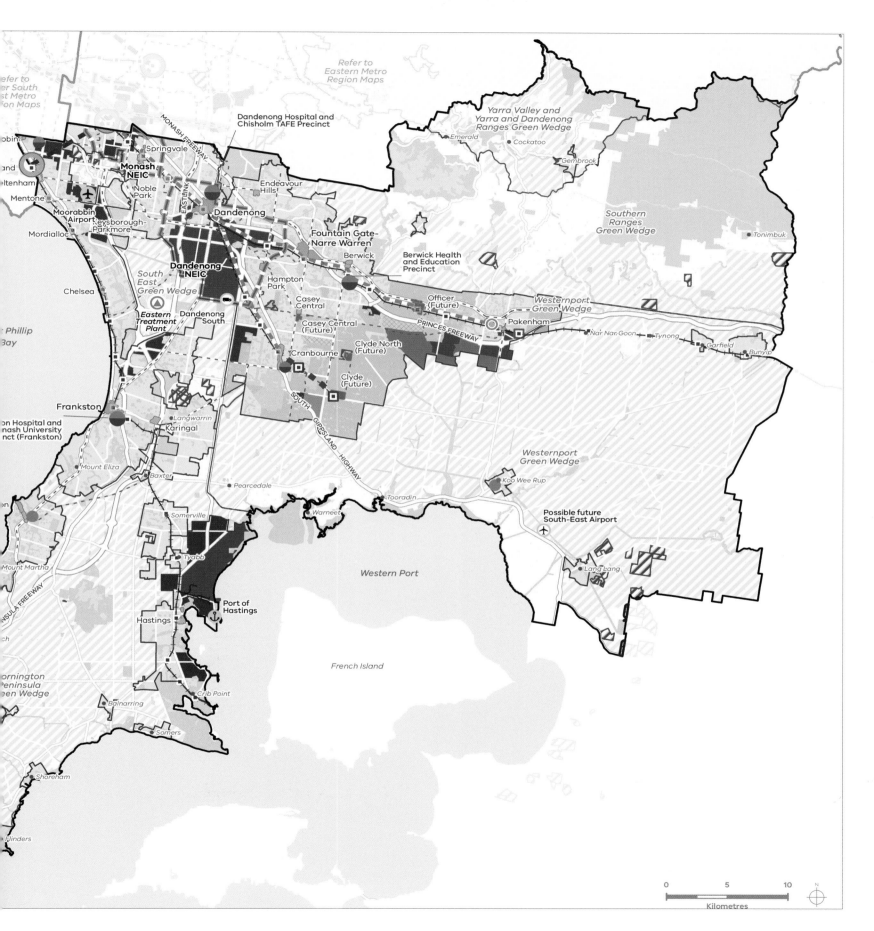

Refer to
Eastern Metro
Region Maps

Refer to
er South
st Metro
on Maps

MONASH FREEWAY

Dandenong Hospital and
Chisholm TAFE Precinct

Springvale

Monash
NEIC

Noble
Park

Endeavour
Hills

EASTLINK

Dandenong

Moorabbin
Airport

eysborough-
Parkmore

Mordialloc

Fountain Gate-
Narre Warren

Dandenong
NEIC

Berwick

Berwick Health
and Education
Precinct

Hampton
Park

South
East
Green Wedge

Casey
Central

Officer
(Future)

Westernport
Green Wedge

Chelsea

Eastern
Treatment
Plant

Dandenong
South

Casey Central
(Future)

PRINCES FREEWAY

Pakenham

Nar Nar Goon

Tynong

r Phillip
Bay

Clyde North
(Future)

Cranbourne

Garfield

Bunyip

Clyde
(Future)

Frankston

SOUTH GIPPSLAND HIGHWAY

Langwarrin

Karingal

n Hospital and
nash University
nct (Frankston)

Westernport
Green Wedge

Mount Eliza

Baxter

Pearcedale

Koo Wee Rup

Possible future
South-East Airport

Somerville

Warneet

Tooradin

Mount Martha

Tyabb

Lang Lang

Western Port

NSULA FREEWAY

Port of
Hastings

Hastings

French Island

ornington
Peninsula
een Wedge

Balnarring

Crib Point

Shoreham

Somers

Flinders

Yarra Valley and
Yarra and Dandenong
Ranges Green Wedge

Emerald

Cockatoo

Gembrook

Southern
Ranges
Green Wedge

Tonimbuk

0   5   10

Kilometres

N

CITY OF LAS VEGAS ZONING URBAN MODEL

City of Las Vegas
Las Vegas, Nevada, USA
By Jorge Morteo

CONTACT
Jorge Morteo
jmorteo@lasvegasnevada.gov

SOFTWARE
ArcGIS Pro, ArcGIS CityEngine, ArcGIS Urban, SketchUp 2020

DATA SOURCES
City of Las Vegas Department of Planning

The City of Las Vegas Zoning Urban Model was created to allow the visualization of 3D proposed projects and zoning plans within the city's downtown districts and redevelopment areas.

This model enables planners to communicate with developers, residents, and elected officials. It also allows planners to identify concerns at the earliest possible time, obtain public feedback, and perform form-based code zoning analysis on parcels proposed for development. Viewshed analysis has been performed at meetings and workshops for residents concerned with obstructed views. Shadow analysis shows the amount of sunlight that is obstructed during specific times of day or of the year in a designated area.

Courtesy of City of Las Vegas.

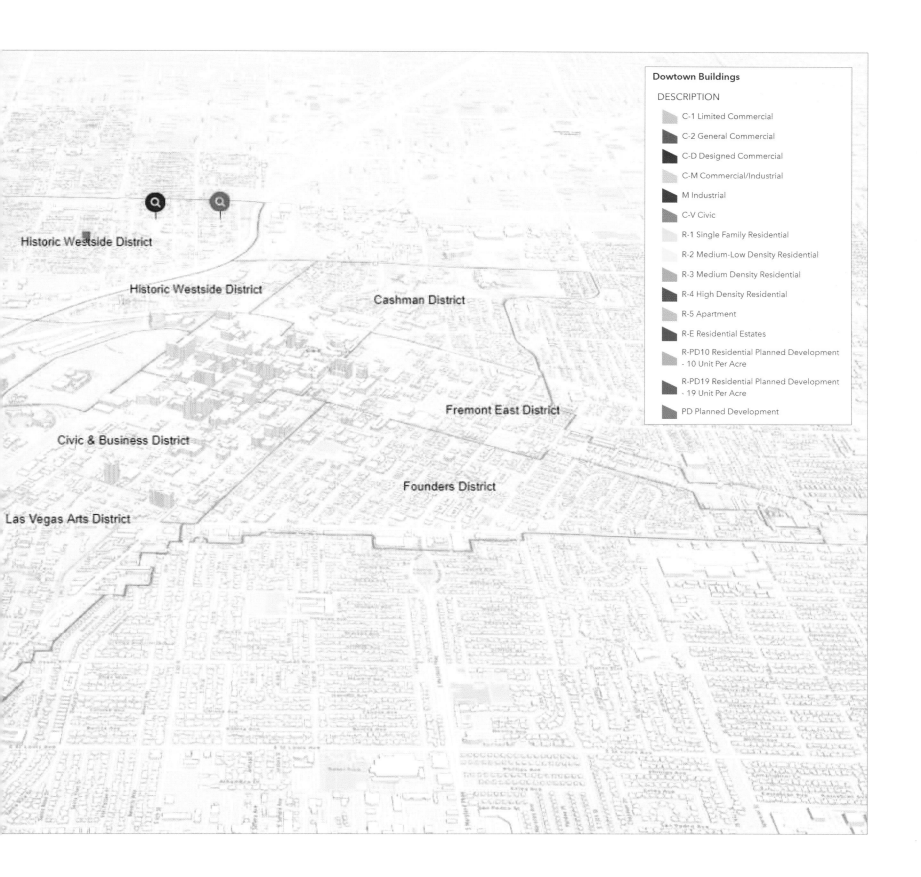

Dowtown Buildings

DESCRIPTION

C-1 Limited Commercial

C-2 General Commercial

C-D Designed Commercial

C-M Commercial/Industrial

M Industrial

C-V Civic

R-1 Single Family Residential

R-2 Medium-Low Density Residential

R-3 Medium Density Residential

R-4 High Density Residential

R-5 Apartment

R-E Residential Estates

R-PD10 Residential Planned Development - 10 Unit Per Acre

R-PD19 Residential Planned Development - 19 Unit Per Acre

PD Planned Development

Historic Westside District

Historic Westside District

Cashman District

Fremont East District

Civic & Business District

Founders District

Las Vegas Arts District

TRANSFORMING UNDERUSED COMMERCIAL PROPERTY INTO HOUSING

Arcadis US Inc.
Belo Horizonte, Brazil
By Clara Freitas, Gabrielle Rossini, Larissa Santos, Luiza Fernandes, Thayná Rombach, and Tamara Montes

CONTACT
Luiza Cintra Fernandes
luiza.fernandes@arcadis.com

SOFTWARE
ArcGIS Pro, ArcGIS Online, ArcGIS API for JavaScript

DATA SOURCES
GeoSampa

The city of São Paulo, Brazil, has a housing deficit of about 1.16 million, but the downtown area has a significant number of underused commercial buildings. The HousingMatch platform, which promotes urban transformation, targets this disparity. In this innovative program, underused commercial buildings are eligible to receive a new housing use designation by the city. Owners of empty spaces and buildings can then connect with potential investors who are interested in social housing programs. Finally, building retrofits adapt the spaces for residential needs, providing vulnerable social groups with dignified and sustainable housing.

This 3D scene shows the buildings and land that are underused or vacant and estimates the potential for new housing units. Users can interactively filter the buildings by type and housing unit and add additional layers as needed. A solution for a better-housed São Paulo, HousingMatch promotes a more inclusive and efficient city.

Courtesy of Arcadis US Inc.

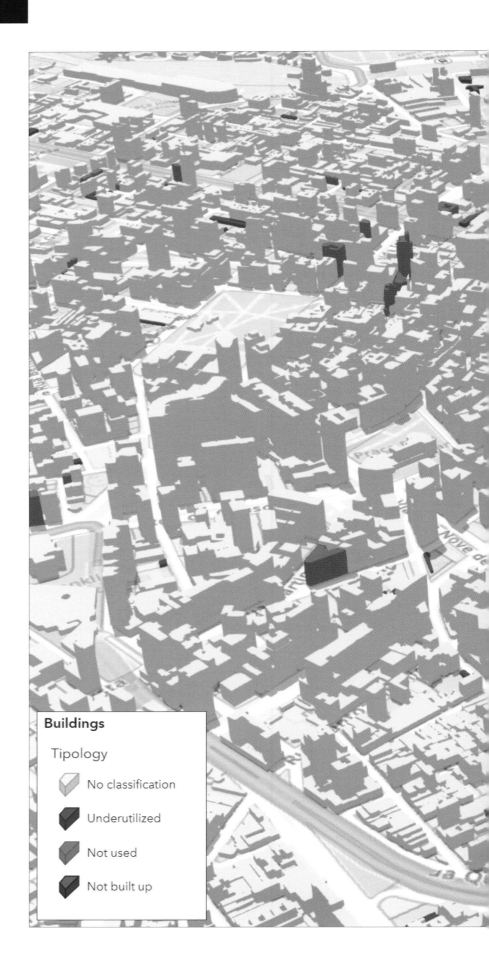

Buildings

Tipology

No classification

Underutilized

Not used

Not built up

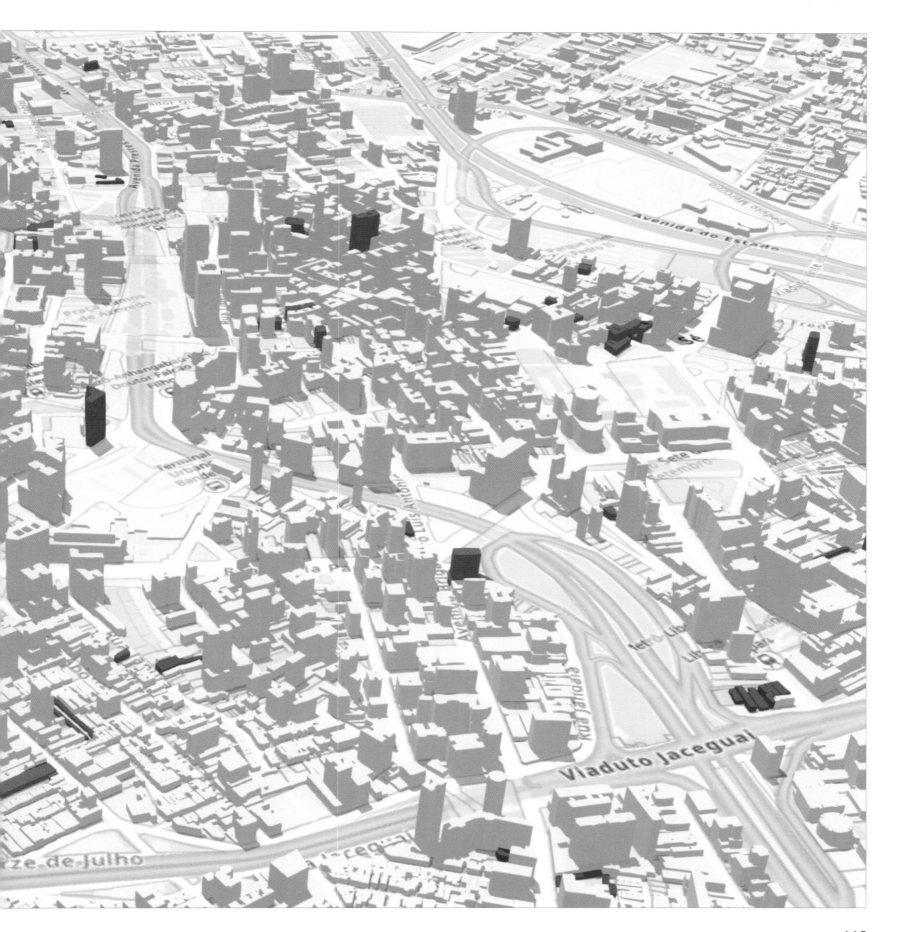

CLARENCE PERRY'S NEIGHBORHOOD UNIT MODEL

Houseal Lavigne Associates
Chicago, Illinois, USA
By Devin Lavigne, Daniel Tse, and Chris Murphy

CONTACT
Devin Lavigne
dlavigne@hlplanning.com

SOFTWARE
ArcGIS Pro, ArcGIS CityEngine, ArcGIS Network Analyst,
TwinMotion

DATA SOURCES
Perry, Clarence Arthur. 1929. *The Neighborhood Unit:
Regional Plan of New York and Its Environs*, monograph I.

Developed in the early 1900s by the architect Clarence Perry,
the neighborhood unit concept was intended to create a
family-friendly, walkable neighborhood. It was developed
in response to the automobile boom in the 1920s and the
subsequent expansion of highways. This map provides an
overview of the concept and illustrates the neighborhood unit
principle in a fully rendered model.

Courtesy of Houseal Lavigne Associates.

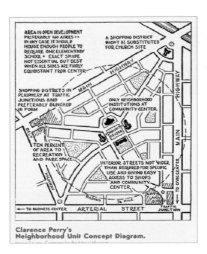

Clarence Perry's
Neighborhood Unit Concept Diagram.

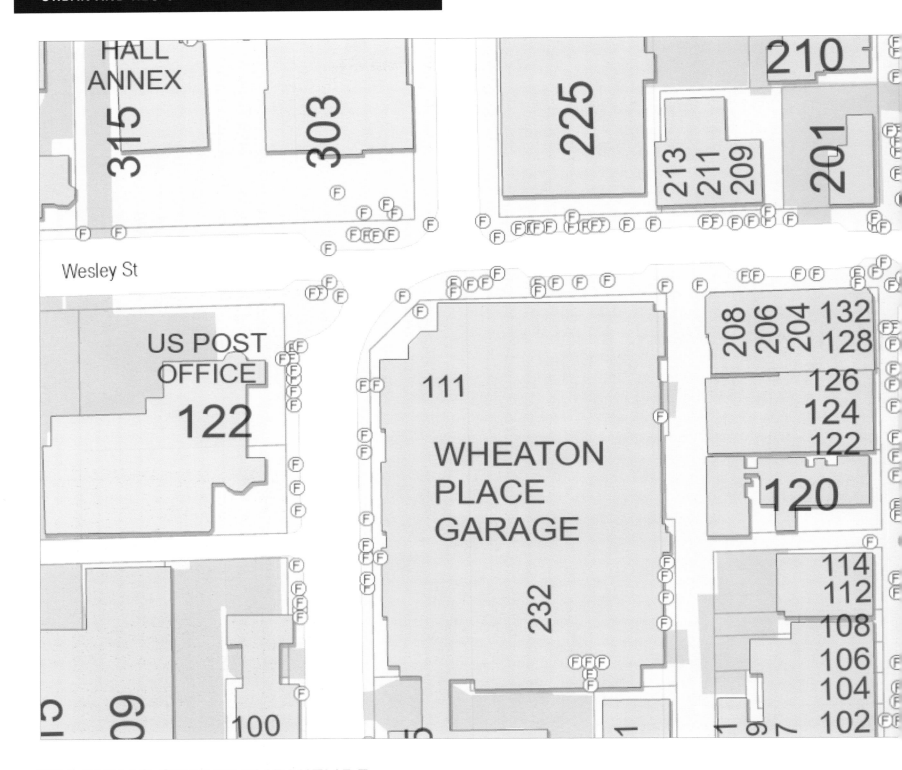

TRACKING STREET FURNITURE

City of Wheaton
Wheaton, Illinois, USA
By Keith Darby

The City of Wheaton, Illinois, is nearing the completion of a $35 million multiyear renovation of the central business district. As part of this project, city employees were asked to provide an inventory of all street furniture assets—trash cans, planters, bike racks, etc. This dashboard shows the locations of more than 1,400 street furniture assets, whose total value exceeded $1 million. The public works department will use this dashboard to track these assets, plan maintenance, and budget for long-term replacement.

Courtesy of City of Wheaton.

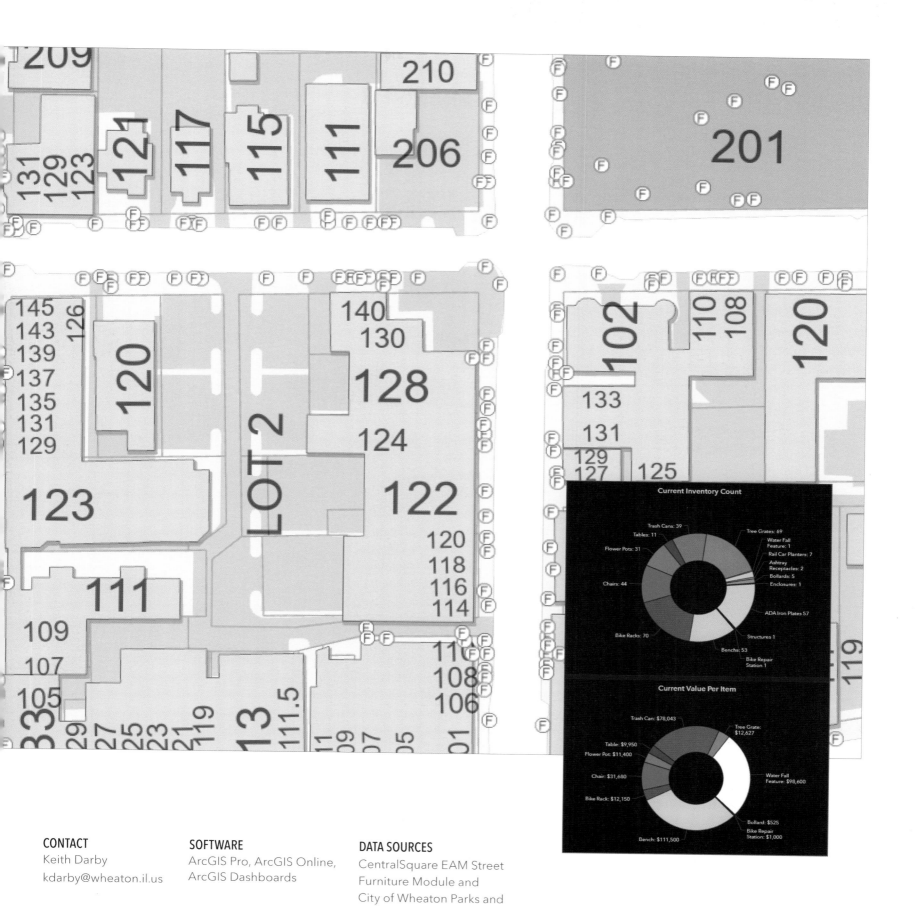

CONTACT
Keith Darby
kdarby@wheaton.il.us

SOFTWARE
ArcGIS Pro, ArcGIS Online,
ArcGIS Dashboards

DATA SOURCES
CentralSquare EAM Street
Furniture Module and
City of Wheaton Parks and
Grounds Division

VISUALIZING THE WAYS BUILDINGS ARE USED

Westport, New Zealand.

Eagle Technology
Westport City and Palmerston North City,
New Zealand
By Omid Khazaeian

The 3D basemap on the left is the foundation of 3D Urban Digital Twin. As opposed to traditional 2D maps, a 3D scene of towns and cities is a more realistic way of visual presentation. In a 3D basemap, more details can be presented in a self-explanatory and user-friendly way, which makes it easier for planners and engineers to monitor assets and for the public to have a better sense of space.

Following New Zealand policy, city councils

must assess and report housing and business development capacity. The interactive 3D scene on the right helps to fulfill that requirement by calculating relevant metrics and visualizing business units as color-coded space use types. By integrating zoning plans, imagery, and business details, this scene can also be used by the public to understand district plans and give feedback to the council.

Courtesy of Eagle Technology.

Palmerston North City, New Zealand

CONTACT
Omid Khazaeian
omid_khazaeian@eagle.co.nz

SOFTWARE
ArcGIS Pro,
ArcGIS Online,
ArcGIS Urban

DATA SOURCES
New Zealand Provincial Growth Fund lidar data (3D basemap),
Palmerston North Pictometry, Esri (3D basemap), Palmerston North
District Plan, Palmerston North Business and Industrial Survey,
(left map); New Zealand Provincial Growth Fund Lidar, Buller District
Water services map (right map)

GENERALIZING WATER RESOURCES FOR EASY COMMUNICATION

US Army Corps of Engineers
Albuquerque, New Mexico, USA
By Douglas Walther

CONTACT
Douglas Walther
douglas.e.walther@usace.army.mil

SOFTWARE
ArcGIS Pro

DATA SOURCES
US Army Corps of Engineers, Albuquerque District

This map portrays the connectivity and relationships of major rivers and dams under the jurisdiction of the Albuquerque District of the US Army Corps of Engineers. It was developed using a schematic style, like a subway map, which allows for the communication of water resource information with management and with employees who are unfamiliar with the district. Relative shapes and locations were preserved, but geographic accuracy was sacrificed to make communication simpler.

Water resources were placed in the foreground, whereas supporting context was moved to the background. Hydrographic basin boundaries show which regions contribute water to each river. Colors differentiate individual rivers, and line width indicates relative importance. Dam symbol sizes and colors indicate management responsibility. The line generalization was developed by locating endpoints and changing direction at the intersections of a regular square grid, and the lines were further smoothed for an improved appearance.

Courtesy of US Army Corps of Engineers.

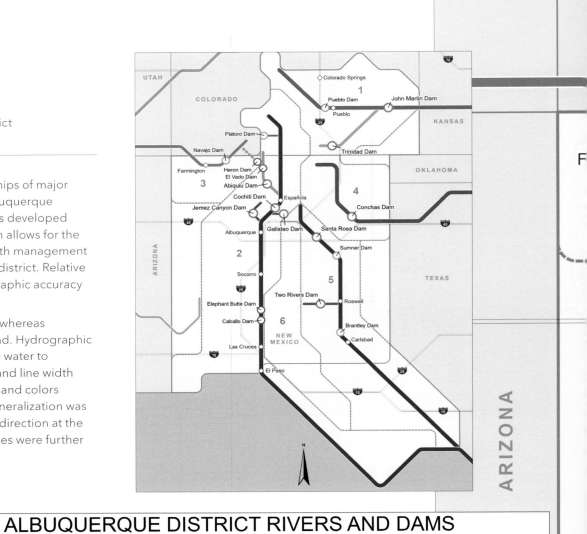

ALBUQUERQUE DISTRICT RIVERS AND DAMS

Symbol		Rivers		Basins
○ USACE Dam	Arkansas River	Pecos River	Purgatoire River	1) Arkansas River Basin
○ Section 7 Dam	Canadian River	Rio Chama	2) Rio Grande Basin
○ BOR Dam	Conejos River	Rio Grande	3) San Juan River Basin
--- Basin Boundary	Colorado River	Rio Hondo	4) Canadian River Basin
— State Boundary	Galisteo River	San Juan Chama Project Tunnel	5) Pecos River Basin
▢ District Boundary	Gila River	San Juan River	6) Tularosa Closed Basin
	Jemez River		

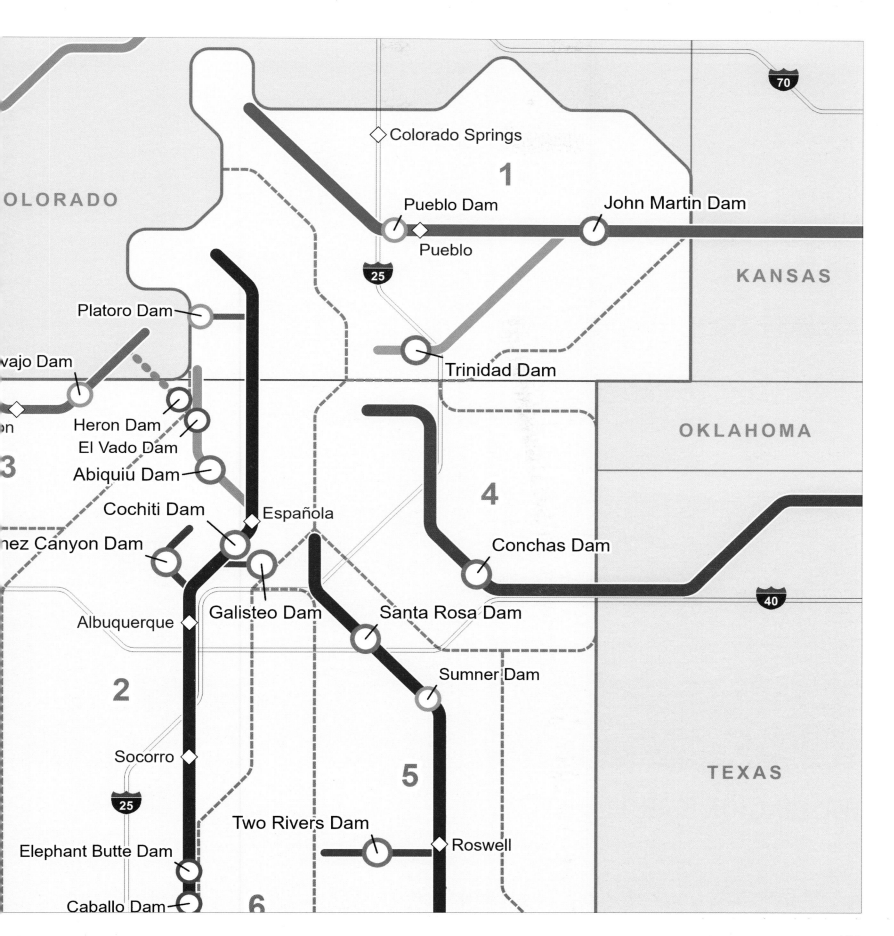

Colorado Springs

OLORADO

1

Pueblo Dam

John Martin Dam

Pueblo

KANSAS

Platoro Dam

Trinidad Dam

vajo Dam

OKLAHOMA

Heron Dam

El Vado Dam

on

3

Abiquiu Dam

4

Cochiti Dam

Española

Conchas Dam

nez Canyon Dam

Galisteo Dam

Santa Rosa Dam

Albuquerque

40

Sumner Dam

2

Socorro

5

25

Two Rivers Dam

TEXAS

Elephant Butte Dam

Roswell

Caballo Dam

6

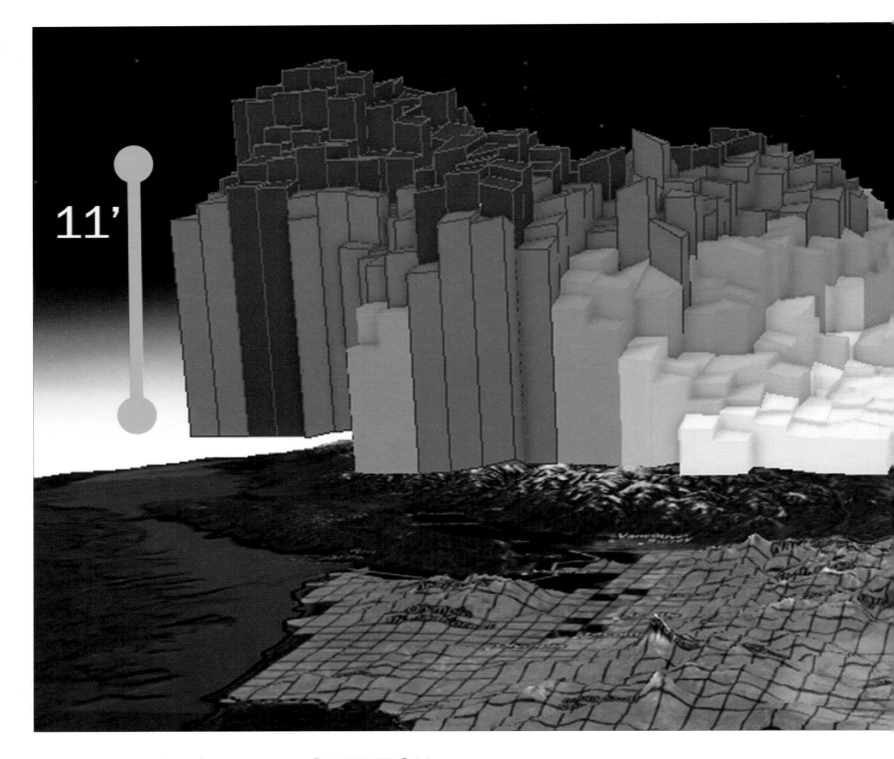

11'

WASHINGTON STATE PRECIPITATION

Washington State
Department of Ecology
Lacey, Washington, USA
By Joshua Greenberg

This visualization of precipitation in the state of Washington uses a floating 3D grid to show the amount that each location receives. Annual rainfall varies by location from 0.5 feet to 11 feet, totaling more than five trillion gallons of rainfall statewide. Mapping this water cycle aids the department in protecting, preserving, and enhancing this precious freshwater system.

Courtesy of Washington State Department of Ecology

0.5'

CONTACT
Joshua Greenberg
josg461@ecy.wa.gov

SOFTWARE
ArcGIS Pro

DATA SOURCES
ArcGIS Living Atlas of the
World

WASTEWATER SYSTEM MAPPING

EPCOR Water USA Inc.
Anthem, Arizona, USA
By Sam Hildebrand

CONTACT
Sam Hildebrand
SHildebrand@epcor.com

SOFTWARE
ArcGIS Pro

DATA SOURCES
EPCOR, Maricopa County (Arizona), Esri

This wastewater system map gives operations, engineering, and planning staff an understanding of how their system functions. Counts show cumulative customer use, and arrows show the overall flow toward the treatment plant. Thicker lines show larger sewer trunks and are color-coded to indicate the contributing wastewater basin.

Courtesy of EPCOR Water USA Inc.

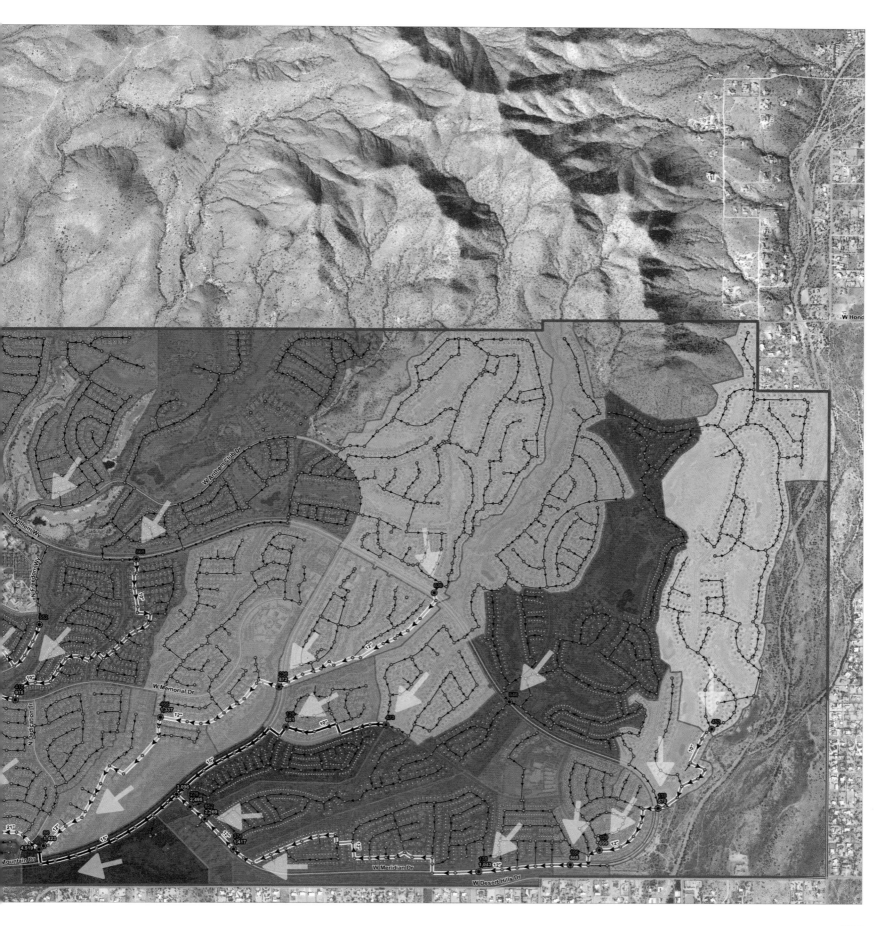

MEKONG DAM MONITOR

Blue Raster LLC, Arlington, Virginia, USA
By Blue Raster

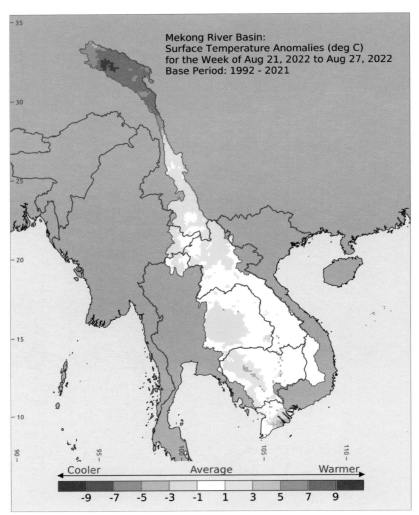

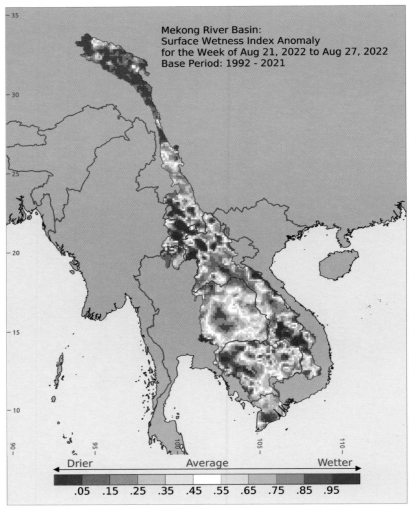

The Mekong River is the world's most productive freshwater fishery, and its natural resources sustain the livelihoods of tens of millions of people who live along the river's course. The nontransparent and uncoordinated operations of Mekong dams in China, Laos, Cambodia, Vietnam, and Thailand create water and food security risks. On occasion, some dams suddenly release water and cause extreme flooding downstream, damaging communities and agriculture.

The Mekong Dam Monitor (MDM) online platform uses remote sensing, satellite imagery, and GIS analysis to provide near-real-time reporting to vulnerable communities and governments in the Mekong Basin to reduce risks related to dam impacts. Before the platform's launch in December 2020, the knowledge of how these dams alter river flow was unavailable. In its two years of operation, the MDM has achieved significant social and policy impact, including social media products, early warning alerts, reports, public events, and workshops, all of which improve accountability, reduce risk, and protect livelihoods.

Courtesy of Blue Raster LLC

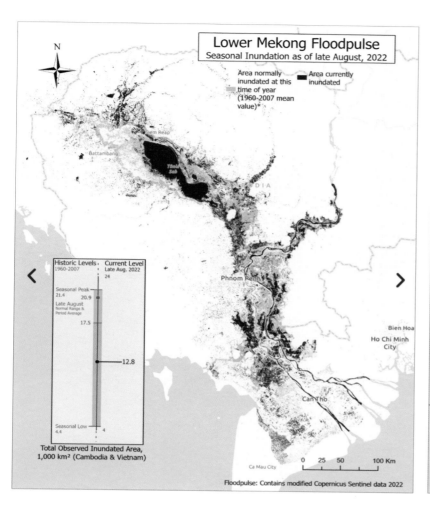

Lower Mekong Floodpulse
Seasonal Inundation as of late August, 2022

Area normally inundated at this time of year (1960-2007 mean value)*   Area currently inundated

Historic Levels 1960-2007 / **Current Level** Late Aug. 2022

24

Seasonal Peak 21.4   20.9

Late August Normal Range & Period Average

17.5

12.8

Seasonal Low 4.4   4

Total Observed Inundated Area, 1,000 km² (Cambodia & Vietnam)

0   25   50   100 Km

Floodpulse: Contains modified Copernicus Sentinel data 2022

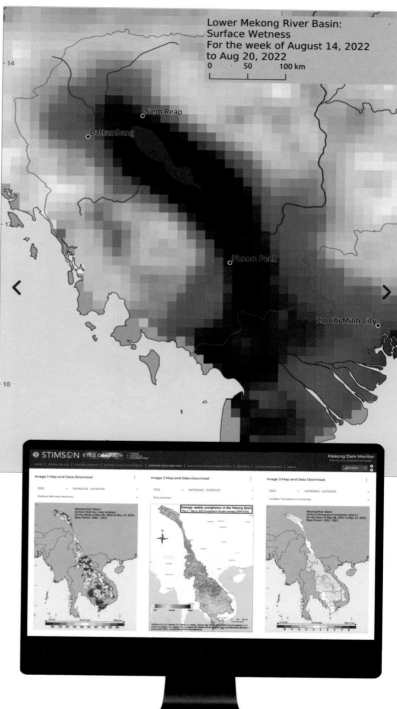

Lower Mekong River Basin: Surface Wetness
For the week of August 14, 2022 to Aug 20, 2022

0   50   100 km

CONTACT
Eric Ashcroft
eashcroft@blueraster.com

SOFTWARE
ArcGIS Pro, ArcGIS Online, ArcGIS Image Server, ArcGIS API for JavaScript

DATA SOURCES
Stimson Center, Eyes on Earth Inc., Mekong Dam Monitor

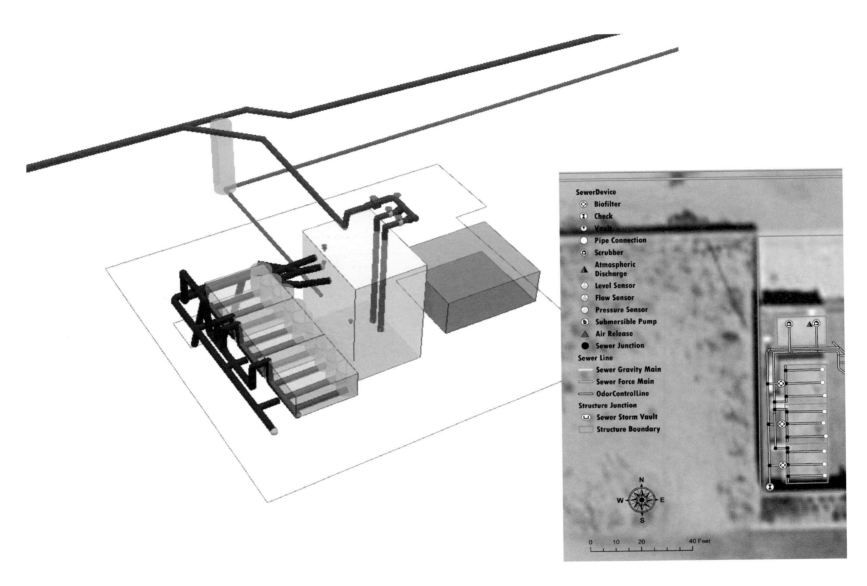

VISUALIZING UTILITY NETWORKS IN 3D

DCSE Inc.
Laguna Hills, California, USA
By Max Ketabi

Historically, utility system visualizations have been done in 2D. However, these 2D visualizations have inherent limitations in accurately representing typical system components: pipes come in varying sizes, valves are complex devices with several components, and manholes are 3D structures with depth and have incoming and outgoing pipes at different elevations. This map was developed to highlight the significant differences between 2D and 3D visualizations, particularly for utility systems.

This 3D model of a sewer lift station shows exactly where network features—such as pipes, cleanouts, and utility access holes—are and what they look like. Users can navigate the scene, identify specific objects, and view related data. Implementing this 3D approach has provided a broader understanding of the site and can be used to train employees in a virtual environment.

Courtesy of DCSE Inc.

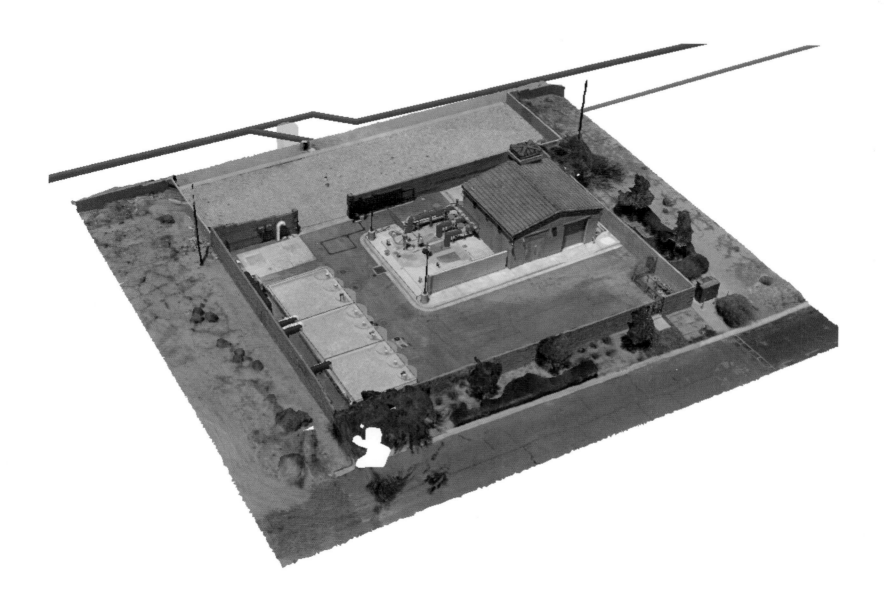

CONTACT
Max Ketabi
mketabi@dcse.com

SOFTWARE
ArcGIS Pro

DATA SOURCES
DCSE Inc., Victor Valley Wastewater
Reclamation Authority

SEWER MAINTENANCE DASHBOARD

City of Wheaton
Wheaton, Illinois, USA
By Keith Darby

CONTACT
Keith Darby
kdarby@wheaton.il.us

SOFTWARE
ArcGIS Pro, ArcGIS Online, ArcGIS Dashboards

DATA SOURCES
CentralSquare EAM Sewer Structure Inspection Module

The Public Works Sewer Division for the City of
Wheaton, Illinois, is responsible for various preventive
maintenance projects for the sanitary sewer and
stormwater systems. The purpose of this dashboard is
to analyze which areas of the city need attention, such as
cleaning catch basins, repairing or replacing structures,
and performing inspections in designated areas. In
this example, the dashboard shows the various sewer
structures that need repairs on the northeast side. The
information is broken down by activity, from routine
repairs to a complete replacement. The dashboard has
helped reduce operational costs and plan for short- and
long-term capital development projects.

Courtesy of the City of Wheaton.

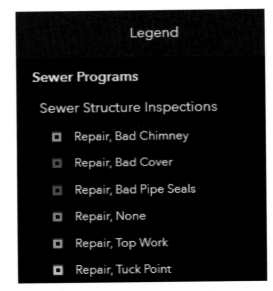

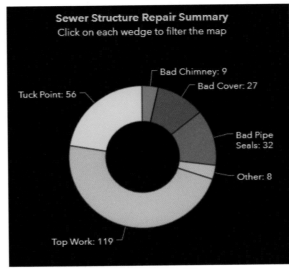

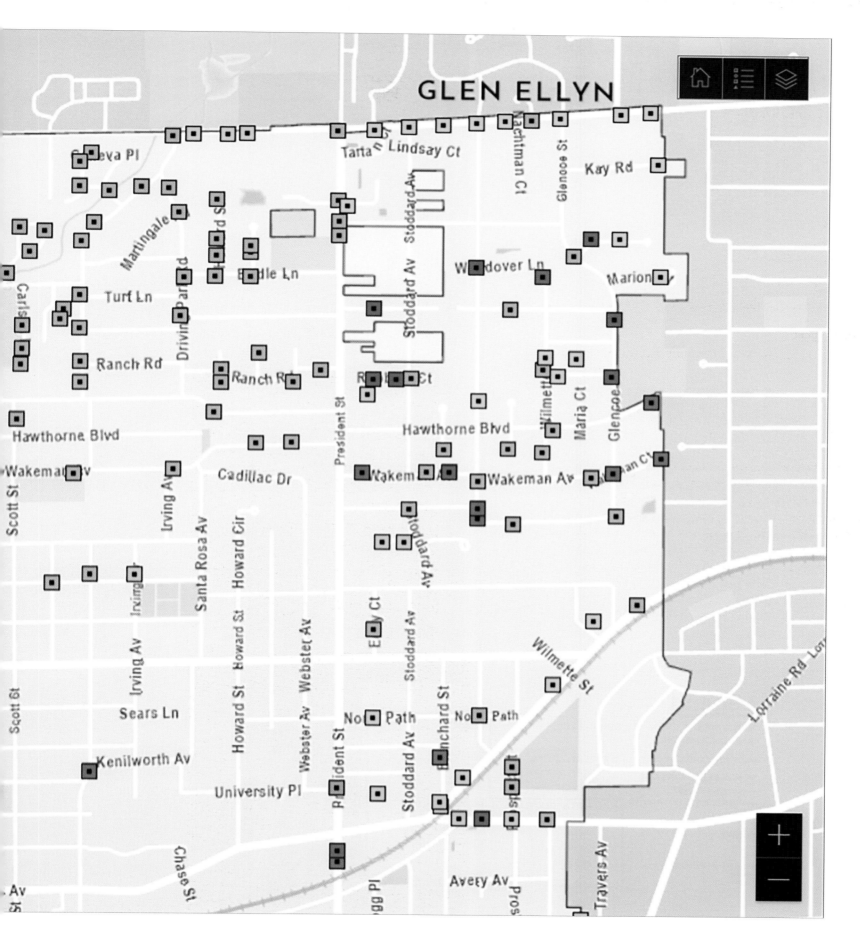

RECLAMATION FACILITY PROCESSES

Eastern Municipal Water District
Temecula, California, USA
By Clayton Gordon

CONTACT
Clayton Gordon
gordonc@emwd.org

SOFTWARE
ArcGIS Pro, Bentley Microstation

DATA SOURCES
NearMap, Loma Linda University, County of Riverside, California State Parks, Esri, HERE Technologies, Garmin, SafeGraph, FAO, Ministry of Economy, Trade, and Industry (METI), NASA, USGS, BLM, EPA, NPS

This reference map of the Temecula Valley Regional Water Reclamation Facility shows GIS features for all structures, roads, land-use areas, and flow directions. It effectively summarizes the wastewater processes for recycling and reclaiming water. The imagery and labels give a clear picture of the ponds, basins, tanks, and mechanical systems.

Courtesy of Eastern Municipal Water District.

DIGESTER

DIGESTER

SITE PLAN AREA 9
SEE SHEET 01C09, 01Y08

EMERGENCY STORAGE POND

EMERGENCY STORAGE POND

Legend

MAIN LINES
⇢⇢⇢ SEWER MAIN LINES
➤➤➤ RECYCLED MAIN LINES

PLANT FLOW ARROWS
→⇢ PLANT INFLUENT
→⇢ TO RECYCLED DISTRIBUTION
⇢⇢ RAW

ABOVE GROUND PIPES
•—•—•

PLANT BOUNDARY
▭

SOLAR FACILITY
☐

SITE PLAN AREAS
⌐ ¬ SITE PLAN AREA
⌐ ¬ SITE PLAN AREA - PILOT

BUILDING STRUCTURE
PLANT 1 (27)
PLANT 2 (11)
PLANT 3 (11)
PLANTS 1 & 2 (13)

BASIN
PLANT 1 (47)
PLANT 2 (34)
PLANT 3 (17)
PLANTS 1 & 2 (22)

TANK
PLANT 1 (6)
PLANT 3 (1)
PLANTS 1 & 2 (4)

NOTE:
(X)=FEATURE COUNT
ALL OTHER COLORS ARE COMMON TO ALL FACILITIES

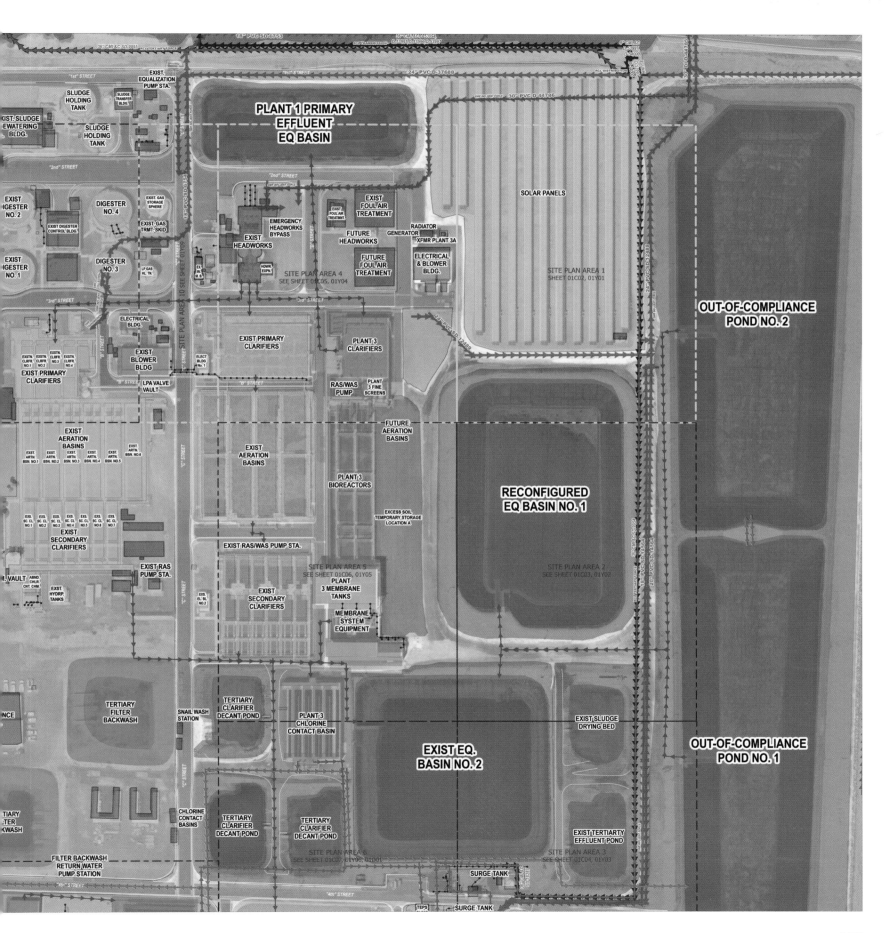

WATERSHED STUDIES FOR MADISON, WISCONSIN

Stantec
Madison, Wisconsin, USA
By Jeremy Del Prete

CONTACT
Jeremy Del Prete
jeremy.delprete@stantec.com

SOFTWARE
ArcGIS Pro

DATA SOURCES
City of Madison, USGS NHD, Esri

Madison, Wisconsin, like many communities, has seen a recent increase in flooding. In response to past flooding damage and the potential for future flooding owing to global warming, the City of Madison has developed models and plans for watersheds of greatest concern in and around Madison. Together, the Starkweather and Olbrich Gardens watersheds constitute the largest and perhaps most complex study to date. The city hired Stantec to conduct advanced watershed modeling and planning services. Modeling will show the extent of flooding inundation levels that can be expected for various types of storms and will inform the future design of green infrastructure and capital improvement projects to mitigate flooding.

This map depicts the boundary of the watershed modeling study. It was used in public meetings so that residents with nontechnical backgrounds could visualize the extent of the watershed and see how elevations vary throughout the study area. The processing of local digital elevation data helped bring the watershed to life.

Courtesy of Stantec.

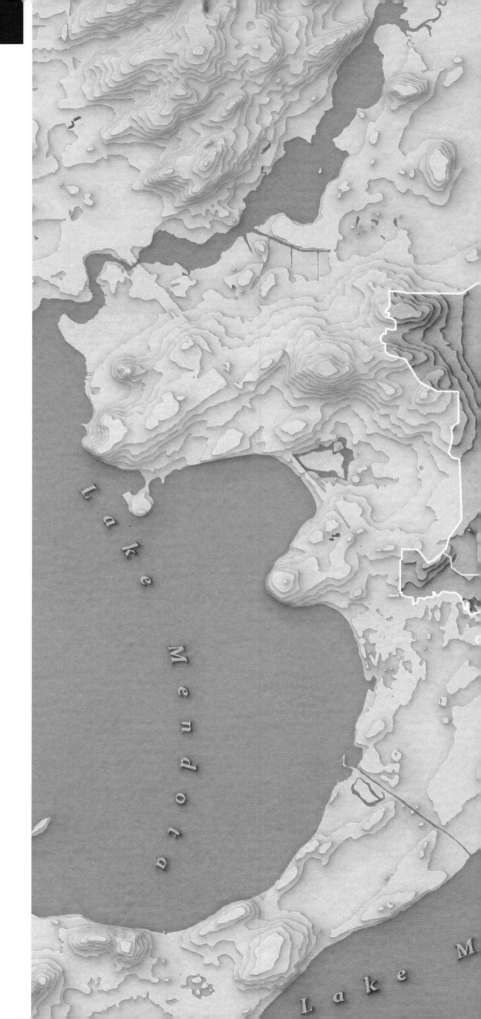

Starkweather Creek North

weather Creek West

Starkweather Creek East

Starkweather Creek Central

Olbrich
Gardens

FLOODING HAZARDS BELOWGROUND

American States Utility Services Inc.
Fort Riley, Kansas, USA
By Joshua Rhamy

CONTACT
Joshua Rhamy
joshua.rhamy@asusinc.com

SOFTWARE
ArcGIS Pro

DATA SOURCES
FEMA. NOAA. Kansas University (consultation with Jude Kasten)

River basin flooding can be a powerful and destructive force above ground, but it can also damage utility infrastructure below ground. This map displays color-coded flood stages in and around Fort Riley, Kansas. Each zone was calculated based on a combination of hydraulic models and other flooding and flow processes, which were then overlaid on a local elevation model. By superimposing this information over utility infrastructure systems in the area, planners can predict which portions of the systems are most likely to be affected. Utility workers can deploy countermeasures to minimize damage to water and wastewater systems and ultimately keep these systems running during a disaster.

Courtesy of American States Utility Services Inc.